Praise for Tony Brown

"Tony Brown is a genuine historic figure—and he is going to continue to make history."

—Newt Gingrich

"Outside the mainstream, outside the Beltway, and outside the box, Tony Brown is a fresh thinker, a truth-seeker, and a genuine visionary. Listen to him—his fingers are firmly on the pulse of what we must know and do to deal with the future coming toward us."

—Dan Burstein, author of *Yen!*

"Tony Brown is one of the most provocative and cutting-edge thinkers in America. His new book is riveting; it will grab you and shake you into reality. At that point you will either do something to improve America or shut up!"

—George Fraser, author of *Success Runs in Our Race*

"Tony Brown has been preaching a gospel of racial uplift through better computing."

—*Newsweek*

"In contrast to the legalistic, civil rights–oriented traditions of the Black establishment, African-American leaders as different in temperament and approach as Booker T. Washington, Marcus Garvey, Malcolm X, Louis Farrakhan, and Tony Brown have emphasized the importance of developing a more self-affirming, economically and intellectually self-sufficient social culture as the primary means of overcoming racial oppression."

—JOEL KOTKIN, author of *Tribes*

"Among the national television celebrities and commentators, Tony Brown is the only one who has questioned the invincibility of the AIDS epidemic. Tony Brown has asked all the crucial questions. In this role as investigative journalist, he has accepted the challenge of those questions and has continued to ask them even when they proved to be politically incorrect. When you read *Black Lies, White Lies*, you will discover that Tony Brown is the scientist among the many celebrities of television reporting."

—DR. PETER H. DUESBERG, professor of molecular biology, University of
California at Berkeley

"When it comes to cultural diversity, Tony Brown does, indeed, speak the truth. Cultural diversity is more than just the right thing to do. It's good for business, good for employees, and it's a real asset to shareholders. *Black Lies, White Lies* is must reading for businesses seeking an enlightened path toward the future."

—JAMES PRESTON, C.E.O., Avon Products Inc.

"I think that Tony Brown is the only one who is advocating in any real way for entrepreneurship in the Black community. He's been very, very vigorous in advocating small business development."

—FRED BEAUFORD, editor of *The Crisis* magazine

"Tony Brown is a tireless crusader who is constantly investigating and examining issues that are vital to the African-American community."

—PRESTON J. EDWARDS, SR., publisher of *The Black Collegian*

"Tony Brown is a wonderful role model for African-Americans. When he speaks, you know that something important will be said."

—RUDY WASHINGTON, executive director, Black Coaches Association

"Tony Brown will make you think and want to act. America needs more of Tony Brown's thinking."

—CLAUDIA J. BOWERS, chairman of the board, Sales and Marketing Executives International

"When I first learned from Tony Brown that he was in the final stages of *Black Lies, White Lies*, I was gripped with great anticipation. Based upon the courageous insights he has consistently exposed for twenty-five years on *Tony Brown's Journal*, I know the significance of this book will transcend America."

—DR. KHALID ABDULLAH TARIQ AL-MANSOUR, author of *Betrayal by Any Other Name*

"One of the little-known aspects of the life of Tony Brown is the high esteem and respect that is extended to him by the black leadership of our nation and of the world. *Black Lies, White Lies* illustrates the reasons for that esteem and respect."

—REV. DR. CHARLES E. BLAKE, West Angeles Church of God in Christ

BLACK LIES, WHITE LIES

BLACK LIES,

The
Truth
According to

Tony Brown

QUILL
WILLIAM MORROW
NEW YORK

WHITE LIES

Library of Congress Cataloging-in-Publication Data
Brown, Tony.
  Black lies, white lies : the truth according to Tony
Brown.—1st ed.
     p.   cm.
  ISBN 0-688-15131-0
  1. Racism—United States.   2. United States—Race relations.   3. Afro-Americans—Civil rights.   I. Title.
  E185.615.B74   1995
  305.89'6073—dc20   95-13315
                                                  CIP

Printed in the United States of America

First Quill Edition

1   2   3   4   5   6   7   8   9   10

BOOK DESIGN BY DEBORAH KERNER

To the two angels,
Elizabeth "Mama" Sanford
and her daughter, Mabel Holmes,
who gave me life when I was two months old

ACKNOWLEDGMENTS

I have learned from this first-book experience that no one person writes a book. The encouragement, patience, compassion, understanding, and love of others are as essential to the completion of a book as the process of creating and writing it.

I thank my sisters, Billie Brown and Jackie McCullough, for their love, encouragement, and concern, and George Fraser, author of *Success Runs in Our Race* and my "networking friend," for introducing me to literary agent Barbara Lowenstein. The late Dr. H. Naylor Fitzhugh, my mentor, smoothed many of my rough edges over the years and taught me the difference between money and wealth, and the real meaning of affirmative action as a spiritual concept and how it applies to the bottom line. He also planted the seeds of many of my philosophies about the value of spirituality and community.

On the technical side, William Morrow, my publisher, allowed me the editorial freedom to say what I believe. My editor, Paul Bresnick, had enough respect for Black people to permit me to use "Black" as the proper noun that it is. He insisted, however, on a readable book that had coherence. Free-lance editor Wes Smith helped me find my "voice" while reviewing and editing my manuscript. The adage that two heads are better than one—even if one is a "cabbage head"—often proved to be true. We never decided which was the cabbage head.

This book represents two arduous years out of the lives of three people who joined with me almost twenty years ago in pursuit of the truth. Technically, they are my employees, but in reality, they are the closest thing I have to a family— those people who worry with you in bad times and share true insights into what the good ones really mean. I lack the ability to tell you how much they have done for me as a person and a friend—or for the completion of this work.

They are three of the most intelligent people I have ever had the pleasure of knowing. But it is their integrity, honesty, compassion, and love that have shaped my appreciation of them. Their loyalty is unheard of in today's cynical world. No monetary value can be placed on their moral virtues.

I speak of Karen Smith, my assistant, who knows and typed every word of this book—and deciphered my middle-of-the-night notes better than anyone else in the world; and James and Sheryl Cannady, the husband-and-wife producer team of *Tony Brown's Journal*. Karen, Jim, and Sheryl spent countless hours as my "family editors," guiding me and bringing me back to reality when necessary. Most of all, they were honest with me. Jim and I worked out the "AIDS" puzzle and tracked the microbes in the two chapters on the threats from drugs and the biomedical establishment itself. Sheryl was better than anyone at continuity, and at detecting the nuances of excess and arrogance that came sometimes with my assumption that I was on the right side. Karen's greatest assets are her profound common sense and her dedication.

But most of all I want to thank God, without whom nothing is possible. As I repeat before every public utterance, my last words before you read this book are:

May God grant me the words to speak His thoughts.

CONTENTS

INTRODUCTION

Before you read this book, I want to warn that you are about to be confronted with truths that may startle and upset you.

You may not like what I have to say, but I guarantee, whether you are Black or White, it will challenge you to review your thinking on vital matters ranging from this nation's economic survival to the failures of Black and White leadership, the origins and causes of "AIDS," the possibility of racial genocide, the potential disasters posed by biomedical research, and the realities of racial politics in this country.

In polite circles, I have been called an "out-of-the-box thinker." A friend offers an earthier epithet: He calls me an "equal-opportunity ass kicker." On the other hand, because I gave up my political independence and became a registered Republican, some choose to see me as a "Black conservative."

In reading this book, you may form your own opinion. I see myself as an American who cares about this country and all of the people in it. Most of all, I care about truth. My journalism has been a search for that truth. I detest lies of any kind from any source, whether those lies are perpetrated by Black elitists who censor anyone who dares to challenge them, or by White liberals who have stifled Black economic development for generations by leading Blacks down the path of dependence. I have no tolerance of racists or demagogues of any color.

I am a Black man whose highly critical views of the Black establishment, the Democratic Party, and White liberals may surprise you. I am no less critical of the White establishment, the Republican Party, and White conservatives. Although I am unsparing in my assessment of the Black community, I love Black people, but I believe Blacks present a serious obstacle to the stability of this nation because too many are still waiting for White people to solve their problems. I believe Black people should let White people go, and solve their own problems.

In this book, which my critics will surely view as a "ferocious polemic," I will write of things that normally are not brought up by Blacks in the presence of Whites. I will discuss the betrayal of the Black community by its misguided, elitist leadership. I will explain why Black Americans are the least successful sociological group in this country. I will also broach a normally taboo topic—the pervasive fear among Blacks that they will become the victims of racial genocide. I will explain why that fear is more grounded in reality than perhaps most Blacks and Whites realize.

The societal decline so obvious today in Black America will afflict the entire nation tomorrow. Drugs, shattered families, teenage pregnancies, and street gangs were Black America's problems yesterday. Today, White suburbanites are afflicted with the same critical problems. The truth is that Black and White, our fates are intertwined as Americans. We have no choice but to put aside our differences and join together to save this country from the sort of economic and moral bankruptcy that has turned once-thriving urban Black neighborhoods into cesspools of violence, drugs, and degradation.

We do not have to like each other. We do not have to live next door to each other. But we do have to share the burden of saving this country and transforming it into a prosperous nation where anyone can grab the brass ring.

In this book, I will be frank about the crises that threaten Black and White Americans and the very existence of this great country. And I will share my realistic and workable ideas for solving those problems, not as Blacks or Whites, but as a team of Americans.

In these pages, I critically examine affirmative action and I am unsparing in my assessment of Black leaders for their policies of the past, but it is essential that this book's central message of self-reliance not be held hostage. I have seen many other Black writers and independent thinkers marginalized for daring to tell the truth as they see it. In this book, I will not limit myself to the issues traditionally deemed appropriate for Black writers.

I challenge everyone who reads this book to look past stereotypes of all kinds, and to clearly see the Black lies, the White lies, and the truth.

CHAPTER 1

DIFFERENT SHIP, SAME BOAT

We didn't all come over on the same ship, but we're all in the same boat.
—WHITNEY M. YOUNG

In the early 1950s, the townspeople of Charleston, West Virginia, practiced a rather benign form of segregation compared to many parts of the country. The White population in my hometown was not consciously intent on keeping Blacks out, but they did like to keep us at a fair distance.

My experiences as a child were therefore rather mixed when it came to racial matters. Before junior high school, my best friend was a White child, my neighbor, Corky. But when it came to the opposite sex, it seemed to me that the prettiest girls in the world were those from Washington Manor, the Black housing project. The great equalizer in my boyhood world was poverty. Nearly everyone I knew, White

and Black, was poor, though Blacks were made poorer by the
fact that welfare and public assistance were for Whites only
in that time and place. The only "welfare" Blacks were allot-
ted was a sack of potatoes, usually rotten, from a federal
program now and then.

The lives of Blacks in Charleston were hardscrabble, but
racism and poverty inspired resourcefulness. There was a
sense that the world owed you nothing, and even if it did,
it wasn't going to pay up soon. So you learned to take care
of yourself and your own family, which was often no small
task. Families were often extended to include cousins, sec-
ond cousins, half brothers and sisters, and even people out-
side of your particular gene pool. Households too were often
organized along untraditional and occasionally shifting lines.

My brother, two sisters, and I were raised separately,
and for a time, none of us lived with either natural parent.
In my younger days, I was reared by two women, Elizabeth
"Mama" Sanford and her daughter, Mabel Holmes. From the
age of two months, when I was delivered to these guardian
angels, they called me "Sonny Boy" because they said I
shone as brightly as the sun.

I lived with these surrogate mothers until they died
within a year of each other just as I reached my teenage
years. I was then returned to my birth mother, who by that
time had also reclaimed my brother and my older sister. My
siblings and I may not have been raised from birth by our
natural mother, but we certainly had been stamped with her
physical features, and we seemed to inherit also her quick
mind and strong will. The key to my overachieving nature,
I suspect, is the nurturing I received from my "mama," Eliz-
abeth Sanford, my supportive schoolteachers, and an ex-
tended family that in many ways encompassed most of the
Black community.

In Charleston, Blacks felt responsible for each other and for each other's children. Adult supervision was considered a community responsibility as well as a parental one. Black children who misbehaved on public transportation could expect to be set straight not just by their own guardians, but by any other Black adults who happened to be present. Our Black schoolteachers also imparted their intense interest in our proper development. In this communal setting, I learned that when you break the rules, you risk forfeiting your standing in the group. This was peer pressure in its most benevolent form, as a force for good. I was taught that in a world that set us apart because of the shade of our skin, we are all we've got, and if we don't stick together, we are doomed.

Early Awakenings

I grew up, then, instilled with the belief that I was under a moral obligation to my community and society in general. At an early age, I developed a finely tuned sense of moral outrage at injustice and dishonesty. The truth, I was taught, could not be denied. My general environment was fundamentalist and proud—to the point of arrogance. If there were Whites who didn't want us around, we thought entirely too much of ourselves to be bothered about it. God would get them for what they did to us, I was taught. Still, when I was a young boy, there were periods when I couldn't wait for God to wade in. I came to see racism as immoral as well as illegal, especially in a nation that preached the Ten Commandments as a way of life and proclaimed that all men were created equal.

That was how I came to view the world I lived in as a child in Charleston. By age fifteen, I was already instilled

with righteous anger over racial segregation when there
came the opportunity to express it publicly. My poor English
teacher, my beloved Mrs. Ruth Norman, did not see it com-
ing when she asked me to represent the youth of my com-
munity as part of the program for a dedication of the town's
"colored" YMCA.

When told I was scheduled to speak right before the
guest of honor, I was flattered to be so well regarded by my
teacher and to be granted a position of prominence before the
entire Black community in a forum reserved for dignitaries. I
had heard Paul Robeson speak and sing at a previous meet-
ing. I remembered also the stone-faced White FBI agents who
never joined us in our thundering ovation for Robeson, our
Black champion. Some said Robeson was a Communist, but
I had no fear of him. He was a Black man who was smart
and important, and he could raise the roof with the resonant
power of his soulful voice. He was big-time, and now so was
I, for I was to be on the same circuit as that great and proud
Black man.

Eager to emulate my hero, I had Billie, my twelfth-
grader sister, help me prepare my speech. I did the writing,
she coached my gestures and pauses. I did what Mrs. Nor-
man had instructed: "Organize your thoughts into a succinct
message, put them on three-by-five cards, use psychological
word triggers, read the ideas over and over, practice in the
mirror, project your voice, use your diaphragm, and remem-
ber, we love you and we want you to do well."

When I got to the Garnet High School auditorium that
evening—it was a Sunday and I was wearing my Sunday blue
suit—I met my three buddies, Arthur Fisher, Lewis Smoot,
and Reginald Taylor. I ignored Mrs. Norman's instructions
to report backstage immediately. I was panicked by the sight
of the crowd, the heavy perfumes of the women, and the

serious mood of the men. So I hung with my three giggling buddies in a secluded section of the balcony. Every few minutes, I checked my three-by-five cards—just in case I was discovered and called to speak.

From my retreat, I could see that up on the stage were various town luminaries, many of my teachers, and a very distinguished and self-assured White man. I was taking in this scene when suddenly the call came: "Will Anthony Brown please come to the stage? Is Anthony Brown in the auditorium?" If I had planned a clandestine exit, my buddies eliminated that option. "Here he is!" they announced, delighted at my discomfort. Wearing my bravest face, I surrendered, reported to the stage, and nervously took my seat. I vaguely remember being introduced to the undistinguishable faces in the audience. I had this sense of being up there alone with my three-by-five cards.

Then the words came spilling out of me. These were powerful words that came not from my carefully prepared speech but from my heart. I had the sense that I was hearing someone else speaking boldly, through me.

"Why have we come tonight to celebrate a damp, raggedy old building with hand-me-down, smelly furniture and flat-sided Ping-Pong balls and pool tables that run downhill?" this bold speaker demanded to know. "Why are we pleased with a facility—if you can call it that—lacking even a swimming pool, while the White YMCA has a first-rate pool just two short blocks down Capitol Street? Why are we content to be second-class citizens and to celebrate our second-class role tonight? No, thank you. When we have something to celebrate, let's do it. Until then, let's do what we have to do to have equal facilities and equal respect under the law. Thank you and God bless you."

In the next few minutes, I was reborn. The YMCA fundraiser crowd burst into a sustained standing ovation. I remember tears welling in the eyes of some of the women, and pride in the set chins of the men, many of whom were high school graduates lucky to get jobs as dishwashers or shoeshine "boys." My own fear that I would be booed or reprimanded by my teachers gave way to a guarded sense of accomplishment. Realizing that I was not going to be reprimanded for speaking my heart, I forgot my fears and basked in the glory, particularly when I saw the reserved Mrs. Norman's enthusiastic approval.

After the audience and the amens quieted, the grave White man took the podium. He said he was moved by my sincerity, and, as diplomatically as possible, he noted that he was in general agreement with my remarks. It would be a waste of time for him to go on any further, he said. So he gave his thanks to the audience and returned to his seat. Something had happened to him also, but I was not sure what it was.

The next day in school, Mrs. Norman called me to her desk and said that the White man whose thunder I had stolen was the chairman of the board of the YMCA, and that he wanted me to present the exact same speech to his board members, the most substantial pillars of the community.

I will never forget the smell of the pool at the White YMCA. It hit me before I saw the pool itself. And then when I did see it, I was dazzled. The water was translucent, mesmerizing, the cleanest, most sparkling water I had ever seen. I was guided past the swimming pool, and when we came to the meeting room where the board members had gathered, I was not escorted in. They were eating. I would wait until I was called to speak.

The chairman of the YMCA board came to get me. He

solemnly reminded me that he wanted his members to hear the exact message I had given to the Black community. I stood at the end of a long table with a confidence inspired by the fresh memory of my audience's reaction at the Black YMCA ceremony. And I let loose with the truth as I saw it upon the White board members. I challenged them to open up our city to Black children as well as White. It would take all of us working together to make Charleston great, I told them.

When I finished this time, there was no sustained standing ovation. Only silence and stares. I had expected at least a small, polite applause. Nothing. Obviously, I had not inspired them as I had my previous audience. It was apparent from their response that the White board members wanted nothing to do with embracing me as part of their community, and nothing to do with me either as a partner in civic development. Clearly, these silent, unapproving men preferred that Blacks remain on the margins of their society.

In the ensuing years, I would see my hometown lose scores of doctors, engineers, lawyers, and other Black human capital as a result of this White ostracism. They drove us away, and we let them.

I received no lunch and no applause at the White Y. I did get a valuable life lesson. In my separate speeches to the Black and White audiences, I discovered the power of truth. Black people, who may have been too intimidated at the time to say it themselves, responded to the truth when they heard it boldly declared, even by a teenager. My speech to the White civic guardians of Charleston taught me that speaking the truth frees you to live without the incumbrance of lies and deceit.

The town's White elite did not like what they heard out

of my precocious tenth-grade mouth, but they were better off for hearing it. In my gut, I somehow understood that in spite of the stoic reception I received from the White board members, I had at least dented the armor of prejudice with the sheer power of the truth. Their own basic sense of fairness had been dulled by generations of hate, but it was still there—and, I believe, it still is present in all people.

A National Dilemma in Black and White

I believe in the inherent goodness of this country and all of its people. But I fear we are on the path of self-destruction as a nation; that the United States of America is committing national suicide. I do not want to see this country destroyed because we fail to see the truth about where we are headed.

My goal in these pages is to alert the reader to the moral and economic crises that threaten to destroy us. By offering the truth, I seek to revive our national courage and our will to survive.

Unless America confronts its racism, its greed, and its moral rot, we face at the very least a drastically reduced standard of living. At the worst, I fear a racial conflagration and national bankruptcy. To avoid these catastrophes and to ensure economic growth, Blacks and Whites must join together to work for the common good on a national scale. We must have the courage to accept mutual responsibility, and to demand change and shared sacrifice from all Americans. If we fail to unite, there will be no Black or White winners, just American losers.

We are all responsible, but Black Americans have an

especially significant role to play in this process of national renewal. Blacks must stop waiting for Whites to rescue them. They must take charge of their own economic development. It is imperative that Blacks take that responsibility and that they become economically competitive through their own initiative. White people are not going to do it for them. It is time we set White people—and ourselves—free of that expectation.

The fate of all Americans, however, is ultimately in the hands of the nation's majority and its ruling class, White Americans. If this nation is to survive, Whites will have to lead us all away from greed and self-interest. Whites must also reject the lie that their fate is not tied to the lives of other racial groups. Whites must come to see the truth in the philosophy of Whitney Young, former president of the National Urban League, who said, "We didn't all come over on the same ship, but we're all in the same boat."

I do not blame Whites for the decline of this country. I believe we are all equally at fault for the present dire circumstance. We all must scrutinize the actions of our leaders and the way those leaders operate. In particular, we must honestly evaluate the failed policies of traditional Black leadership and those of its liberal White supporters who have made Black America ashamed of itself and dependent on handouts for survival.

This book, then, is not about the problems of Black Americans and what Whites must do about them. It is about the problems of all Americans, and what we all must do to resolve those problems.

Socioeconomic Metastasis

The primary threats to our national sovereignty are the loss of moral virtue in the American character, racial conflict between Blacks and Whites, and the burden of national debt. These are not new problems, but they have become malignant and now demand our focused attention. These three threats have put enormous pressure on our society, and pressure seeks relief indiscriminately. It is up to us to determine whether America's response to the growing societal pressure is positive action or negative reaction.

These threats endanger us all, but currently, they weigh most heavily upon Black Americans, whose underclass is the most exposed and vulnerable segment of society. There is an old saying among Black people: "When Whites catch a cold, Blacks get pneumonia." As in any community, the weakest fall first to an epidemic; eventually, though, even the strong fall victim. Drugs, crime, moral degeneracy, the devaluation of life—we all face these societal threats because a malignant cancer spreads rapidly through our entire culture. I call this process *socioeconomic metastasis*.

It has been observed also that when the more successful and affluent White population feels threatened or comes under pressure—for example, when a recession lowers the standard of living—Blacks become scapegoats. History is full of episodes in which the dominant culture group (such as the Aryan Germans) takes out its frustration on a numerical minority group (such as the Jews), by first marginalizing them and then attacking them as scapegoats, which in the end did nothing to solve Germany's economic problems. White Americans should do all they can to ensure the physical and economic health of the Black community, not out of philanthropy, but out of their own self-interest.

Symptoms of social disorganization that manifest first in a society's most marginal group also confirm structural shifts that have been taking place over a longer period at the economic level. Deficiencies in the Black community often are the result of the same underlying economic turmoil that will eventually become apparent in the White community. The epidemic of illegitimate births among the "gangsta" culture subset of young Black men and women and among low-income Whites, for example, has now spread to the much less marginal groups—including White middle-class teenagers. Between 1940 and 1950, illegitimate births to teenagers constituted 4 percent of all births. The percentage started going up dramatically every year beginning in 1970, and today over 30 percent of births are to single women.[1] The figure is projected to hit 50 percent by 2003. In some areas and among the most marginalized groups, the illegitimacy rate already far exceeds 50 percent.[2]

While Black women have historically led the rate of out-of-wedlock births (80 percent of Black births are illegitimate; 22 percent of White births are illegitimate), figures now suggest that socioeconomic metastasis has kicked in.[3] Between 1980 and 1989 the rate of increase among Black women was 40.4 percent, compared to a huge 85.6 percent increase for White women.[4] From 1982 to 1992, the rate of illegitimate births increased overall from 15.8 percent to 24.2 percent. That was the largest increase in the history of this country. During that period, illegitimate births increased most for White women.[5]

This is a frightening and unacceptable trend. If the societal impact of unmarried teenage mothers was considered burdensome among a numerical minority group, consider what it will be when the epidemic spreads among those in the middle-class majority.

11

The Spread of Drug Abuse

By looking carefully at the social and economic conditions of the most marginalized group in our society, we can see the future of the most privileged sector. The drug trade, probably the fastest-growing sector of the American economy, also illustrates how the problems that afflict the lowest strata of society eventually spread to the higher levels.

During the Harlem Renaissance of the 1920s, Whites in New York City had to go "uptown" to Harlem to buy cocaine. Many believe that law enforcement authorities realized they could not stop drug use—or perhaps they had no interest in stopping the lucrative illicit business—and so instead it was decided to allow Harlem and Black neighborhoods across the country to flourish as "safe zones" for drug purchases in order to isolate the accompanying problems.

Like vendors at the ballpark, organized crime was awarded the drug concession in Black neighborhoods. And then, after decades of being allowed to build an efficient manufacturing and distribution system in Black neighborhoods, the marketers of illicit drugs branched out. Today in America, Whites no longer have to go uptown to get cocaine. They can find it on their own school playgrounds, their college campuses and office buildings, or even the neighborhood pizza parlor. Several years ago, federal investigators discovered that organized crime was using pizza parlors in small towns across the country to distribute drugs to the White middle class. Investigators found that the "Pizza Connection," as it came to be known, reached into the nation's heartland.

Today, 60 percent of the illegal drugs distributed in the world go to customers in the United States, which has only

4 percent of the world's population. The drug lords can thank racism for allowing their networks to become established. And we can thank it also for the lives destroyed by drugs, not just in Black neighborhoods but across the nation.

Many White Americans still do not realize that by allowing drugs to flourish in the Black community, they poisoned their own sons and daughters. America's socioeconomic crisis is primarily visible as a Black predicament today, but I predict that it will be—just as drugs are today—a national dilemma within the next decade as the middle class is hit with shrinking employment due to technological displacement. Traditional inner-city problems—crime, drugs, poverty, illiteracy, homelessness, welfare dependency, illegitimacy, school and domestic violence, unemployment, school dropouts, gang activity—are all headed for the White middle class. By the year 2000, we may well have a large White underclass with social pathologies identical to those of today's Black and Hispanic underclasses. At that point, with that many citizens impoverished, our nation will be more like a Third World country than America the beautiful.

Little Gangstas on the Prairie

Not too long ago, I told my television viewers that to get a picture of White America tomorrow, all they had to do was take a photograph of Black or Hispanic America today. But in many ways, and in many parts of the country, the metastasis has already become painfully obvious.

Focus, for instance, on the 2,500 members in twenty-five street gangs that have been identified by police in the blue-collar Midwestern burg of Davenport, Iowa, and the surrounding Quad Cities area. Just as Black rhythm and blues music was embraced by young middle-class Whites in the 1950s and 1960s, "gangsta" values are invading the middle-class White mind-set. Criminologist James Houston told a *Spin* magazine reporter who examined the phenomenon of gangs in the Corn Belt that teenage Whites are attracted to the gang mentality and the allure of sex and drugs in part because of "overall economic hardships."[6] When both parents are forced to work, when financial pressures disrupt the family and threaten security, self-esteem drops. And when self-esteem is at low levels, gang membership thrives. Gang leaders, who use their members as sacrificial lambs, prey upon the need of young people to belong and to be recognized. Lynn Kindred of the Davenport police department described gang-banging in the heartland: "These little white shitheads act black, they talk black, they think they're tough, but they are mutts without their gangs, just idiots. They get recruited by the older black members, who run them like tops."[7]

One seventeen-year-old Davenport gangsta told *Spin* that he is a "W.A.R. lord"—a "White African Relative." He added that he did not like "straight white honkies. I like to kick their asses."[8] Midwestern White youths threatening "straight white honkies" is almost too bizarre to believe. Although it has been difficult for many small-town Midwesterners to accept the fact that their sons and daughters are Black-gang-member wannabes, one incident brought the reality home to the Whites in the Quad Cities area.

A seventeen-year-old White girl was partying late one summer night with boys from her school when they apparently hatched a plan to rob a convenience store in order to finance their budding cocaine operation. When asked to lend her car for the robbery, the girl refused. Police say she was then raped, beaten, and shot through the back of the head with a shotgun. Her body was dumped on a rural road. Within fifteen hours, six teenagers were in custody for the crime.

"The first day after the murder we didn't have any pictures of the arrested," *Quad-City Times* reporter Cheri Bustos told *Spin*. "I know everyone around here thought, 'Oh. These black kids again.' Then we ran the mug shots, and I can tell you there was real shock. There were six white teenage faces staring off the front page."[9]

As the gang situation in Davenport illustrates, while poor Blacks generally have been the first to fall victim to societal problems and perils, we are all increasingly vulnerable. Senator Daniel Moynihan has estimated that eight of ten Black children will be paupers before they are eighteen years old.[10] I predict the same will happen to White children within two decades, if current trends continue.

In the 1960s, Moynihan accurately predicted the breakdown of the Black family unit. Based on the theory of socioeconomic metastasis, I am predicting the breakdown of the nuclear White family by the end of this decade—poor Whites first, swiftly followed by the middle class. What kind of America will we have when the majority population, a base of approximately 200 million people, has a poverty rate of 32 percent (equivalent to that of Blacks in 1994) and one out of every three Whites is living under Depression conditions? Or when eight of ten White children become paupers before they become eighteen years old? Or when street gang values overwhelm traditional values across society?

Malignant Economics

Annual budget deficits and our national debt weaken the very foundation of our national economy. If Americans can address critical economic problems, especially the debt, we might have a much greater chance to avoid self-destruction as a world power. Economic stability would not only help the underclass where the short-term adverse results are already visible, but spare the far greater majority of Americans who eventually would be affected.

Everyone in America is linked economically. As the prominent economist Lester Thurow pointed out in *The Zero-Sum Solution*, those in the upper-income classes should realize the inherent danger to themselves in the present decline of the middle class. The "hammering" of the middle class, Dr. Thurow writes, will "move on" to hit the upper classes as immediate economic problems and in the long term as social chaos. "Today production jobs moving offshore; tomorrow engineering, design, and managerial jobs are apt to be moving offshore,"[11] he has said.

As an illustration, Thurow offers the General Motors–Toyota joint production facility in California where autoworkers have regained their jobs (the cars will be built in America), but all of the engineering and design work is done in Japan. "If this arrangement foreshadows the future, America regains some middle-income jobs but loses upper-income jobs," Thurow notes. "Similarly the 13 million video recorders being purchased by Americans but built by the Japanese represent the loss of a lot of engineering, design, and management jobs. What is today a threat to the middle class will tomorrow become a threat to the upper classes."[12]

We can already see the impact that foreign competition,

a high-tech workplace, downsizing, restructuring, and enormous accumulated national debt are having on the general population and relations between population groups. As productivity in this country declines and growth slows, group rivalries are heightened. When we feel powerless, we blame one another. For example, it is generally assumed by many Whites that Blacks and affirmative action are mostly responsible for the recent decline of wage parity of White males. The true reason is a structural change in the economy that has more to do with globalization and technology than with Blacks and affirmative action. As Thurow explained to me, "In the 1980s, White male high school graduates were the big losers. This was because they kind of had a monopoly on the automobile, the machine tool, the steel jobs. Those are the ones we lost to the Japanese, the Koreans, and the Europeans."[13]

Teamwork

Like it or not, Blacks and Whites are in this together. We win as a team or we lose as a team. Since we are all Americans, we play on Team America, and, frankly, our team is in the dumps right now. The decline is obvious on all socioeconomic levels, and it is rooted in our loss of moral character. Instead of looking for ways to help this country prosper as earlier generations did, we look to prosper from government help. Parasitic lobbyists, who poison the political process, are the most blatant of those in line for handouts. These and other white-collar parasites suck the blood out of the institutions and values that hold the democratic system together.

The corrupting influences of lobbyists and political action committees and special-interest groups seem more deeply damaging to me than young inner-city men who sell drugs for enormous short-term financial rewards even as their comrades are ritually sacrificed every night on the televised news.

What the average hardworking American does not seem to understand is that the problems in the underclass cannot be arrested, imprisoned, and then forgotten. What happens in Watts or Liberty City or Harlem is merely a harbinger of the future for your neighborhood and mine. Therefore, it is imperative that we halt the rapid deterioration of our nation's economic and moral systems if we are to protect all levels of the socioeconomic strata of society. I believe we can do that with a program that offers opportunities to those who show both potential and real need for assistance in developing that potential. I arrived at this concept after pondering an unusual source—Christ's Parable of the Prodigal Son.

America's Prodigal Sons and Daughters

In my opinion, all Americans could benefit from this lesson, agnostics and atheists included. The Parable of the Prodigal Son serves as a blueprint for breaking the cycle of racial animosity and relieving our economic crisis. It offers a lesson of love and compassion and fairness. The Prodigal Son is considered the "gospel within the gospel" by theologians, and to me it has additional meaning regarding the role of the individual in the nation and the nation within the individual.

In this famous parable, Jesus taught his disciples about a man who had two sons. The younger son asked his father to give him his share of the family wealth so that he could go off on his own. The father complied, dividing the property between the two sons. The younger son quickly ran off and squandered his share in sinful living. Near starvation, he decided to return home and beg for forgiveness: ". . . But when he was yet a great way off, his father saw him, and had compassion, and ran, and fell on his neck, and kissed him. And the son said unto him, Father, I have sinned against heaven and in thy sight, and am no more worthy to be called thy son."[14]

His father sought no apology, nor did he condemn the wayward son. Instead, he ordered a feast and a celebration: "For this my son was dead, and is alive again; he was lost, and is found."[15] The father ordered the preparation of "the fatted calf." Upon learning that his father was celebrating the return of the younger son, the older brother became angry and questioned his father's judgment. The father answered: "Son, thou art ever with me, and all that I have is thine. It was meet that we should make merry, and be glad: for this thy brother was dead, and is alive again; and was lost, and is found."[16]

My reading of this parable is that the father (government) in his wisdom knew that to help the forlorn son (those in greatest need) is the most pragmatic way to help everyone in the family (nation). To have given a fatted calf (entitlement benefits) to the faithful son (the stable population; the bedrock of society) would not have increased the overall stability or output (goods, services, and personal wealth) of the family unit (society in general). But in strengthening those who are without the basic requirements for successful participation in society—education, child care, health care, training, financial aid, etc.—we strengthen the overall stability of society.

To divert a young person from becoming a drug pusher to becoming a doctor who saves lives rather than kills people saves more than lives. It saves society the collateral damage and the expenses of a destructive and irresponsible life. Society benefits in additional income and productivity. Only in strengthening the weakest link in the chain do we strengthen the entire chain.

The Prodigal Son's rehabilitation was the difference in the growth of the family—not the son whose contribution could already be depended upon. The father's action was as pragmatic as it was charitable. The father was helping the group by helping the most needy son. By rehabilitating the weakest son, the father enriched his family, making it possible to raise more crops and fatten more calves in the future. In the most fundamental sense, it is love, forgiveness, and compassion in a society—its essential fairness—that drive its productivity, its gross domestic product.

If the Prodigal Son analogy is too spiritual for you, let me explain it in hard, cold, capitalistic free-enterprise terms—a more modern example of the sort of affirmative action that makes sense to me: the player draft for professional sports. This form of affirmative action was pointed out to me in one of the many lessons that I received from my mentor Dr. H. Naylor Fitzhugh, who in 1931 became the second Black to receive an M.B.A. from Harvard. At the end of the regular season in professional sports, what team gets first choice at selecting the best new player to be drafted? The answer, of course, is the team with the poorest winning record—the worst team in the league. Why, one may legitimately wonder, would a system allow the worst team in the league to get the best new player? My Prodigal Son logic is that if you strengthen the worst team in the league, you strengthen the overall league. This means that the entire league becomes more competitive, because on any given day or night any team in the league can defeat any other team. A more competitive league increases consumer participation and media attention. The TV networks make more money from the advertisers because more people watch and attend the games, and the team owners increase their earnings; as a result, because the wealth of the league has grown, the players can successfully bargain for larger salaries.

The spectators win, the TV networks win, the owners win, and the players win. But most of all, the public wins. It is a win-win situation—only because we strengthened the weakest team in the league. Now that's affirmative action.

However, the proponents of *privileged-class* affirmative action reject that need-based emphasis. They argue for preferences for Blacks, Hispanics, women, Asians, and Native Americans, among others, without any distinction between the classes within those respective groups. Could it be that there are Blacks, Hispanics, women, Asians, and Native Americans who are perfectly capable of making it on their own? Too often, our government offers a support system for the son who stayed at home at the expense of the one who left.

The *New York Times* reported that during the last three decades of affirmative action preferences, "the proportion of poorest Blacks has grown."[17] And in a testimony to these elitist preferences, the *Times* attested to affirmative action's inherent inefficacy. Although the already upwardly mobile

Blacks have benefitted enormously, "those programs left legions of blacks behind. . . ."[18] Unarguably, affirmative action has not made a major impact on poverty and is a failed racial remedy. Opponents argue that the beneficiaries of affirmative action are mostly members of the African American middle class who do not need the aid of the program anyway. Elitist supporters of upper-class preferences counter by arguing that they have never claimed affirmative action to be a self-sufficient policy that will address social and economic disadvantage; instead, it should be applied at "gateway points" for the middle class.

This "Me-ism" argument is a tacit admission that privileged-class affirmative action is concerned neither with moving people from poverty to the middle class nor with overcoming the past effects of discrimination. Instead, privileged-class affirmative action is applied mainly at the "gateway points," in employment and higher education to make the entry smoother for the handful of inner-city refugees who by some miraculous set of circumstances reached the "gateway points" or those middle-class, upper-class, and wealthy Blacks who have outperformed 90 percent of the White population educationally and financially. In fact, middle-class White women are the chief beneficiaries of affirmative action. And one-fourth of the contracts awarded by the Small Business Administration as set-asides in 1994 went to 1 percent of the "minority" firms. This classic privileged-class affirmative action concept is snobbish, uninformed, nonproductive, and divisive.

My characterization of this affirmative action approach is that of a bait-and-switch scam. If young Blacks and Hispanics overcome racism and miraculously score 1200 on the Scholastic Aptitude Test, then, and only then, privileged-class affirmative action will sponsor them as targeted "achievers." But, as we have seen far too many times, privileged-class affirmative action is not designed to help these ethnic minority academic achievers overcome the formidable barriers confronting them before they reach the "gateway point."

Blacks graduating from Harvard and other elite schools will not be the salvation of the Black community. Too many of them choose not to reinvest their talents into the Black community anyway. Their ambition is all self-directed. The focus should not be on saving these who qualified for the most elite schools—these high achievers would undoubtedly have made it one way or another. Instead, we need to concentrate on those who might, with some assistance, qualify for less elitist campuses. For example, every time you take a drug pusher off the corner in Harlem or a youth out of the cotton patch in the rural South, you change and improve the character of the Black community. The Black graduate at Harvard, or even Howard, was going to graduate from somewhere anyway. The pusher and cotton picker had no other options. Because of my socioeconomic status, if I received an affirmative action transfer benefit, I would defer it to a socially or economically disadvantaged person living in Appalachia or Harlem. That would do the most good for all of us.

Giving Help Where It Is Needed the Most

My proposal for improving our failed affirmative action policy is quite simple. I propose instead a universal *Affirmative Opportunity Program*, a modified and refocused form of affirmative action. Simply stated, it is designed to do something good for someone who truly needs the help and is ready to seize opportunity. Blacks will benefit disproportionately because they are disproportionately in the underclass, and Whites will dominate as numerical beneficiaries. But needy people would be helped. Thus, it becomes a *mainstream* opportunity program to rebuild the nation, rather than as a payoff (albeit for a noble reason) to the middle class in various aggrieved groups. The program, in this manner, becomes an affirmative opportunity rather than another middle- and upper-class entitlement benefit.

How can America as a nation be helped in the long term by providing affirmative action only to middle- and upper-class Blacks, Hispanics, Native Americans, women, and rich Asians? And how could these groups be harmed by affirmative opportunity? If 100 percent of the people benefiting from an Affirmative Opportunity Program qualified based on need—whether they were Black or female or Hispanic or White or Native American or Asian—it would not matter, would it? Because the most needy Americans would be getting help—and that would maximize the benefits to all of us.

The most efficient way to develop human resources is suggested in these words from James Russell Lowell:

Not what we give, but what we share—
For the gift without the giver is bare;
Who gives himself with his alms feeds three—
Himself, his hungering neighbor, and me.

That lesson came home to me one Sunday afternoon, April 13, 1992, at the Black Expo U.S.A. at Philadelphia's Civic Center when I told fifteen hundred Black people that the salvation of the Black community is the development of Black people into "human capital." Income and wealth are different, I said. Income is what you earn; wealth is what you're worth. And what you're worth is based on what you know. An informed, educated people will earn large incomes; the uninformed and uneducated will permanently be marginalized groups on society's economic and social fringes. My theme was that when we help the least in our group, we do the most for our group.

The next morning, I found the opportunity to put into practice what I had been preaching. A CHILD SHINES AMID THE SHAMBLES," said a six-column, front-page headline in *The Philadelphia Inquirer*. Reporter Kimberly J. McLarin had written a wrenching story about a young Black girl of great promise who was in danger of never realizing that promise: "Karesha Lowe is 14, and fatherless and poor, with a mother serving life, a half sister as a reluctant guardian, a crowded house, an angry brother and an intellect so hungry she thinks algebra is fun,"[19] said McLarin's opening paragraph.

In one way, Karesha'a life was no different from those of millions of Black youths trapped in a cycle of poverty, war-zone schools, and antisocial gangsta values. In another sense, put eloquently by McLarin, Karesha was "a rose growing through rubble, one that will unfold or be trampled underfoot."[20] Those words burned themselves into my mind. Unless someone helps her, I thought, she will undoubtedly be trampled.

Karesha was not yet eleven when her father was shot to death in a fight. She was barely twelve when her mother was convicted of murder and sentenced to life in prison. The pretty teenager and her thirteen-year-old brother, Chris, were living in a "dim and crowded row house" with a stepsister who had three children and two grandchildren of her own. In spite of all the tragedy in her life, Karesha was making straight A's at Vaux Middle School. She had graduated from eighth grade at the top of her class and was accepted into one of the best public high schools in the city.

In spite of the well-adjusted math whiz's accomplishments, I knew that she was not out of the woods. To succeed and reach her potential, Karesha needed a supportive home life and a college education. Without help, she might not reach her full potential, I realized.

In a series of newspaper columns, I appealed to the readers of 130 Black newspapers across the nation: "These are our children, our potential human capital; this obligation is our responsibility." If the Talented Tenth—the elite of Black society—has $16 billion to spend on annual meetings of Black organizations each year, we certainly have enough money to save the boy or girl who may discover the cure for some mysterious disease, or set up the first space station on Mars, or become the first Black to be elected President of the United States.

I appealed for donations to establish a fund for Karesha's education and maintenance. Contributions poured in, from Blacks and Whites. We raised $6,025 very quickly for clothes, books, food, and incidentals. The funds are administered by me and two of Karesha's former teachers, Lynne Johnson and Florence Johnson, who have become her surrogate mothers and her mentors.

On May 13, almost a month after I first learned of Karesha's plight, I held a press conference at the University City Science Center in Philadelphia. Dr. Isaac B. Horton III awarded a full $50,000 scholarship for Karesha to the College of Wooster, Ohio. It was sponsored by the Delaware Valley chapter of the National Organization for the Professional Advancement of Black Chemical Engineers in conjunction with the College of Wooster, which had responded to my appeal.

In addition to the $50,000 scholarship and the $6,025 in donations, an invaluable present was made by the Muncy State Prison officials, who at my request allowed Karesha's mother, Cheryl Casper, to make a surprise visit at the press conference. Those who donated to this cause would have been paid back tenfold had they witnessed the joy and pride. Karesha wrote to me afterward: "Not only did you help change my life for the better, you also changed my mother's, my sister's and my brother's. . . . And for that I thank you with all of my heart."

Karesha Lowe now attends the exclusive George School in Newtown, Pennsylvania. She still loves math, has served as a member of the student council, and has tutored seventh-graders after school. Teacher Florence Johnson sent me a report on Karesha from George School along with the comment: "All who have invested time and money in her have made a great investment."

I fell short of raising my goal of $100,000 for Karesha, and I have been accused of grandstanding by a handful of middle-class Blacks. One Black elitist offered that I had wasted my time and money because there were so many others who needed help. Her rationalization, and this is what it is, soothes her conscience and simultaneously provides her with a perpetual excuse to do nothing but complain about what White people will not do.

I have received some criticism, but I have no regrets about helping Karesha. I take great pride in the knowledge that Karesha will probably never need public assistance. She will not be a criminal, and she will be less likely to be a victim. Instead, she will be in a position to help make this more the kind of world we want to live in—in fact, she is already working with the disadvantaged, sharing her gifts. And as a successful doctor, chemist, astronaut, or President of the United States, she may remember that someone helped her when she needed it.

Karesha Lowe is the Prodigal Sister. She was lost and now she is found. And in finding her, those of us whom God used to help her helped ourselves, and our nation.

CHAPTER 2

ENTITLEMENT SOCIALISM: WHY AMERICA IS NOT WORKING FOR U.S.

Even the opponents of Socialism are dominated by socialist ideas.
—LUDWIG VON MISES

Sick of greedy politicians with their hands in the public's pockets, newspaper columnist Mike Royko once suggested that the official motto of the City of Chicago be changed from the Latin phrase *Urbs in Horto* or "The Garden City" to *Ubi est mea?*—"Where's mine?"

Chicago didn't adopt the motto of greed, the entire country did. We have become a nation on the dole, a people constantly foraging for government handouts. Americans, rich, poor, and in the middle, have come to feel that they are entitled to tap into the public till. These exorbitant "entitlements" gladly handed out by politicians in exchange for votes now pose the greatest threat to the stability of this country.

29

Contrary to popular misperception, it is not just the poor in the United States who are welfare-dependent. Entitlements are delivered to nearly every segment of the population. Social Security, Medicare, Medicaid, farm supports, veterans' benefits, federal pensions, unemployment compensation, food stamps, and Aid to Families with Dependent Children (AFDC) are but a few of the entitlements handed out. These programs alone account for over half (53.5 percent) of the total U.S. federal budget. Add in administrative costs and tax expenditures and the bill for these entitlements totals $1 trillion every year. The recipients include three-fourths of all American households.[1]

Consider also that three-fourths of our federal entitlement dollars go to people who have not demonstrated financial need. Only one-sixth of government subsidies are allotted to help the impoverished. The middle class and the elderly receive 75 percent of all federal health benefits but constitute only one-eighth of the total population.

Seventy percent of our "glorified welfare system" payouts go to 1 percent of the population—and that includes Social Security recipients who receive every dollar paid during their working years plus interest in just four years. And yet, the poor and those with needy children are hypocritically scapegoated as "welfare queens" and parasites.

David Kline, managing editor of *Critical Intelligence*, is among those alarmed over federal handouts. He and others have wisely come to understand that it makes far more sense to target the fat cats on Social Security than the poor folks on public aid. Kline has noted that only one in eight federal dollars goes to the poor. Households making less than $10,000 received $5,700 in benefits in 1993, but those with incomes over $100,000 received $9,300.

"Out-of-control entitlement programs . . . suck up two-thirds of all federal outlays, but they do not fight poverty," Kline has written. "Cutting welfare and other social services by a whopping 25 percent, for instance, only saves us $5.05 billion over five years—or about $18 and some change per citizen. But then you have to figure in the cost of picking up the bodies off the streets of our major cities every morning and disposing of them—not to mention the added expense of dealing with food riots and the like—so my guess is the net savings would probably be a lot less than $18."[2]

Kline suggests that it would make far more economic sense to impose an "affluence test" that would reduce just slightly the Social Security payout for people with incomes over $40,000. This, he projects, would save more than twenty-six times what a 25 percent cut in welfare and social services would salvage. And, he noted, "It would only affect wealthier individuals who do not really need all those free benefits."[3]

Nearly 25 percent of government benefits go to families with incomes of more than $50,000, according to Peter Peterson, a former U.S. Secretary of Commerce and author of *Facing Up: How to Rescue the Economy from the Burden of Crushing Debt and Restore the American Dream.*

Peterson and others savvy to the real world of government handouts to the nonneedy have been warning us for years that if the United States continues to dole out Social Security and Medicare benefits at the current rate, the present youth generation will very likely become known as "tax slaves."

Let's face it: We all feel we've got it coming, and then we wonder where it's all going. A poll showed that 81 percent of Americans believed health care was a "right" and not a privilege that must be earned. Sixty-six percent felt the same way about housing. A 1993 poll of Black Americans found that 63 percent "believe that the government should guarantee every American a job or a decent standard of living."[4]

In the previous chapter, I told you that I believe that Black and White, rich and poor, are all in the same boat. In this chapter, I am going to tell you how we are ripping the boat apart even as we pack together on it. Unless we stop demanding government handouts, I believe we will all, very shortly, find ourselves clinging to the debris of a democracy ripped apart by collective greed.

The national debt, which is the result of our greedy grab for entitlement programs, will destroy us. Few people understand the debt or its devastating impact. Many politicians depend on this general lack of knowledge.

The debt that is slowly turning all Americans into indentured servants is being caused largely by *entitlement socialism*, which I define as a system in which elected officials use the government treasury to transfer wealth to special-interest groups in exchange for their votes. The public demands federal benefits for its political support—its votes as well as its campaign contributions. At least forty million Americans who donate more than $4 billion a year to twenty thousand special-interest lobbying groups make up the not-so invisible government that controls the political process.[5]

Not only is this a corruption of the democratic process, but, as a result, an oppressive debt burden now threatens to punish all Americans with national bankruptcy. Our personal wealth is not the only thing at risk. Our freedom is endangered too. To put it in concise terms, each American owes about $18,000 toward the national debt, and that amount goes up every second of every day. As a nation, we owe nearly $5 trillion, and we continue to borrow more each year—somewhere between $250 and $400 billion depending on who you talk to. That $250 to $400 billion deficit is the difference between what we spend and what we earn. Living beyond our means has become the American way of life for nearly all Americans, and our leaders.

For years our elected officials and their economic experts have told us that these deficits did not matter, that the red ink only covers a small percentage of gross domestic product. They also have said that if we reduce our federal budget too quickly, we could bring about a recession or even a depression.

We have been borrowing voraciously since 1963, when Lyndon Johnson found it was a convenient way to finance two wars—the War on Poverty and the Vietnam War. Since then, Republican and Democratic Presidents, with the help of their comrades in Congress, have mortgaged our futures, handing out entitlements and pork to us while all the time billing it to the accounts of our children and grandchildren. It has been a convenient way to please the voters. Fiscal caution means nothing to a public whose motto is "Where's mine?" And it certainly means nothing to politicians whose primary concern is staying in office so they can get theirs.

Politicians who have the courage and integrity to offer serious deficit-reduction programs—such as former Representative Timothy Penny of Minnesota, a Democrat who resigned from Congress; Ohio's Republican Representative John Kasich; and Democratic Senator Bob Kerry—have become political suicides in the current milieu. "Should a ceremonial sword be the prize for lawmakers who dare to give voters what voters claim to want?"[6] *Time* magazine asked.

Former U.S. Senator Warren Rudman said that the 1992 presidential election was our last chance to head off fiscal disaster by making the debt a priority. Twenty million Americans heeded his call and voted for Ross Perot, whose primary platform was getting rid of the national debt. Did the rest of the electorate blow it? Though he promised to address the budget deficit, Clinton has done nothing to reduce the debt; instead, his budget only increases it.

Some economists hold that the deficit is even greater than anyone has let on. In an interview shortly after Clinton took office, author and hyperinflation expert Gerald Swanson updated his book *Bankruptcy 1995* by providing three hair-raising facts: (1) The Treasury Department's fiscal year 1992 statement showed that deficit spending was being understated. (2) Treasury reports show that the 1992 deficit was not $290.2 billion as officially reported, but a whopping $403.9 billion. (3) Our government during the first three months of fiscal year 1993 increased the deficit by 43 percent over the same period during fiscal year 1992.[7]

Reagan inherited a $994 billion national debt from Jimmy Carter, and after Bush succeeded him, they left Clinton a $4.4 trillion time bomb. This was the bad news that greeted Clinton when he took office. It had to be a shock for a man who, like his Vice President, has never worked in the private sector or ever had to risk his own money to meet a payroll.

Clinton has tried to convince the public that his budgets have harnessed the annual federal deficit. To the contrary, the deficit in fiscal year '95 siphoned off 50 percent of the net savings of businesses and households. Clinton only plans to trim the deficit by a third (but increase the debt) until 1996, when it will blast off again with even greater velocity.

In fact, the deficit will grow more over the next decade than in the previous one, and the Clinton debt will increase by over $1 trillion.[8]

Hard Times Require National Unity

I believe that within the next few years, ever-expanding national debt could possibly trigger very serious racial polarization, loss of homes, shortages of all kinds, middle-class unemployment, widespread bankruptcies, and a totalitarian form of government. It is time we all recognized this threat and united against it.

Whether they be retirees, militant feminists, Black nationalists, or White reactionaries, special-interest groups should realize that at this crucial juncture in history it is necessary to broaden their focus. Right now, the most important mission is saving our country. Being an American is what is important right now. We have got to come together to save this nation.

Current events remind me of the days preceding Pearl Harbor in 1941. Warning signs were everywhere, but the government and the people ignored them. It was a ritual of denial that also preceded the Civil War. But most Americans are unmindful of history, just as they are unmindful of the future. Only the moment counts.

We are all in the same boat, but we don't have all of our oars in the water. At this moment, America's saving rates are the lowest among the industrialized countries. Real wages and total compensation—the nation's best standards of the quality of life—are failing or stagnant. Our students and workers lag behind in their training and productivity, according to the federal Competitiveness Policy Council, which noted a twenty-year decline in the nation's global competitiveness.

Tax Slaves

In a section of the official 1995 U.S. budget entitled "Generational Accounting," Boston University economist Lawrence Kotlikoff explains how future generations will become slaves of the federal debt. He projects that Americans born after 1992, because of the burden of an aging population, runaway government spending, and a huge national debt, will pay an average net tax rate of 84 percent. The rate was 82 percent when Clinton became President. His projections show that the first generation to get hit by this tax tidal wave will be the baby boomers, born between 1941 and 1961. They can expect to lose benefits and pay a higher tax rate of about 35 percent. To give a comparison, the generation born at the turn of the century has a rate of 23.6 percent.[9]

Kotlikoff calls the federal debt a "generational nightmare" that is the result of our unwillingness to confront "the coming crisis in America's out-of-control entitlement programs." He notes that two entitlement programs alone, Social Security and Medicare, account for almost 40 percent of Clinton's proposed 1995 budget.[10] By the time children born this year reach their mid-forties, the cost of Social Security and Medicare alone will eat up 38 to 53 percent of the average worker's taxable wages.[11]

I like a plan that would base all entitlement programs on need and limit federal assistance to $40,000 per family.

With that one action, we would save 25 percent of the $800 billion budget expenditure for entitlements—or $200 billion a year. That amount could retire the annual deficit in one year.

These and other cuts would be accompanied by a reformed tax system that eliminates the personal income tax and the corporate profit tax that taxes what people put into society. They would be replaced by two consumption taxes—a national sales tax on consumers and a value-added tax on businesses—on what people take out of society. This revision produces a budget surplus (see details in Chapter 12). We could eliminate the debt in just a few years and shift the emphasis to saving instead of consumption, and investing instead of borrowing. It would boost productivity and the nation's standard of living.

My budget proposal is better because it is economically sound and morally truthful. The great political animal Bill Clinton admitted in private that reelection, not responsibility to the people of this country, is his primary concern, according to *Washington Post* editor Bob Woodward, author of *The Agenda: Inside the Clinton White House.* On more than one occasion, according to Woodward, Clinton described his own economic program as a "turkey"—while at the same time he was selling it to the American people as the nation's salvation.[12] This is untruthful and morally evil.

House of Horrors

In his disturbing book, Woodward describes a White House of jealousies, deceptions, raw ambition, and a frantic group of people who want to hang on to power, not to save our nation from its own greed.[13] And certainly not sacrifice and change.

Honest citizens and patriots can no longer get elected to public office because they are unwilling to pander to the voters with payoffs and to destroy the economic structure of the country. We have evolved a system of governance that locks out the solutions and shuns the best leaders. We cannot accomplish politically what is demanded economically because, in truth, we are abusing democracy by courting socialistic tyranny. We want the government to support us. That is the dilemma of America's modern-day democracy. Even now, if we could mobilize enough people as reform agents, the political and economic systems may be so structurally flawed that they would not respond.

I would end the counterproductive dependence on entitlements. The only thing an American is entitled to is freedom; anything else is a gift.

America has been warned silly about its imminent collapse, but Americans are mesmerized by an almost religious faith—a blind faith—in government's ability to support the good life. "We all know that we should consume less and invest more in education, research and development, plants and equipment, and infrastructure. But we don't,"[14] economist Lester Thurow has said.

Despite pious assertions to the contrary, improving the general welfare and leaving a healthy America for future generations are not top priorities. Amazingly, with the death of socialism in most parts of the world, America is rapidly moving in that direction, proving, perhaps, that capitalism is a transition stage to a socialist state, as Karl Marx prophesied. I say that conditionally because the not-too-gradual transition of the American marketplace economy to a socialist state is not a natural evolution; it is the result of a corruption of the free-market system and the destruction of our economic infrastructure through government intervention, spending, and taxes.

The catastrophe of a socialist government coup d'état is not inevitable. If it happens, it will come because Americans are simply too self-centered to change and sacrifice—as were the Romans, Greeks, and Britons. The United States is on the path toward poverty and unremitting pain.

Americans seem to be inviting a government takeover of their lives—which is the essence of socialism. Since World War II, the socialist ideal of the government as a demigod has gradually conquered the consciousness of a critical mass of Americans. The history of socialism does not recommend it for serious intellectual consideration. Its attraction lies in the seductive promise of something-for-nothing—especially when socialistic ideas are marketed under new labels such as "entitlement" and distributed willy-nilly. Millionaires have a fivefold return on their Social Security investments because politicians pander to the elderly vote.

Driving Blindly

How is it that the world's champion of democracy and free markets, the United States, is embracing socialism at a time when we have just witnessed the catastrophic crash of the Soviet Union? It has happened because Americans have little sense of either the past or the future.

One of the biggest offenders, the Reagan administration, gave free-lunch socialism a Republican spin with supply-side economics. Many jobs were produced, but they came at too high a price. Although a good case can be made for using borrowed money to defeat the Soviet threat as Reagan did, we could be only a few years behind the Soviets in realizing our own deep-seated structural problems.

Former *Wall Street Journal* economics editor Alfred Malabre, Jr., wrote in *Beyond Our Means* that since the Reagan White House "the United States [has] had a productivity problem, brought on by a combination of too much borrowing, too little saving, and too much spending." He noted that most of that spending has been on nonproductive areas such as defense, interest on debt, and various entitlements "as opposed to spending that would eventually strengthen the economy, such as for new plant and equipment, education and the like."[15]

The key to restoring our standard of living is to restore our productivity as a nation. And in the immediate short term, the country would need to respond to the federal debt crisis as if it were waging a war for survival, which would be exactly the case. Both actions would require new political leadership, a sense of urgency, perhaps public outrage, and a new national purpose based on unselfishness.

None of this is likely to happen unless the dangers of entitlement socialism are understood by the voters and the leaders they elect. What we need is what we do not have: a President willing to tell the people that the news from the competitiveness and productivity front does not bode well for our future standard of living. Moreover, we do not have a political process that would then support a President in changing what needs to be changed.

In 1992, we elected a new President from a different political party with a fresh generational outlook, but the same inertia grips the process, and the debt continues to escalate. Instead of mobilizing to reduce the debt, we continue on a government-led consumption binge. And because we consume more than we produce, we must borrow more and more. Often, we attempt to justify our high living by claiming that the United States has "the most productive worker in the world."

Lies. The truth is that United States average output per worker per hour of work is in decline relative to the increases of Japan, Germany, and the other Big Seven nations, which will tie the United States between 2000 and 2003 and dramatically move ahead thereafter.[16] Only by raising output can we increase the standard of living. The election of Clueless Clinton would have been a setback at any juncture in history, but at a time as crucial as this—an economic meltdown point—his policies and lack of honesty are a disaster.

Dr. Feelgood Is Out of Anesthesia

I am not claiming that bankruptcy is a possibility; I am saying that I do not see any way to avoid it, unless a lot of prominent economic thinkers are wrong and the voters rebel against entitlement socialism—immediately. A happy ending is possible only if the American voters wake up to their own involvement in this potential disaster. I say "potential" because we can still do something about it.

The 1996 presidential election may give Americans their final chance to choose a leader and members of Congress who will be allowed to tell the truth and get elected because of it. The message: If Americans do not change and sacrifice, the country will decline into a Third World totalitarian state. Term limitations will not work. If voters force politicians to represent the general welfare, rather than the moneyed interests that keep them in office, it would not matter how long they remain in office.

The American voters will not avoid a financial collapse by projecting their behavior onto the politicians and punishing the politicians for the faults of the voting public. Politicians are simply doing what the public tolerates—and ultimately wants. Unfortunately, 43 percent of American voters picked just the man for President in 1992 who could keep the economic train wreck on schedule. The voters turned to another Dr. Feelgood who promised the middle class more benefits than his opponents did. Now as President, his actions ensure the debt will live on to haunt us all.

Race War

Aside from the very real danger of national bankruptcy in this country, the next greatest threat to the United States is a race war. And an economic collapse could well be linked to the tinderboxes of America's inner cities. While the American dream is fading for many Whites as they watch their standard of living and way of life fade, it has never come to fruition for millions of Blacks trapped in the urban underclass. They have been economically and socially marginalized for generations. Many are without hope. I have seen it firsthand.

During the 1968 riots in Detroit, I remember seeing young, unarmed Black men, bare-chested, sweaty, and vacant-eyed, walk into a unit of fully armed paratroopers with fists flailing and take the beating of their lives. Others stood defiantly in front of tanks and did not lose the dare.

I remember an incident during the planning of a peaceful

march that I coordinated which featured Martin Luther King. A White cop pulled me out of a crowd of about thirty jaywalkers on the campus at Wayne State University in Detroit, where I was a student. He upbraided me, as he jerked me across the street by the arm. He was doing what they had done to Blacks for years. On-the-spot justice. It was not that incident alone, being singled out and disrespected because I was the only Black on the street, but it was the accumulation of almost daily harassment—being pulled over on the street and forced to get out of my car as though I were a convicted felon. It was not unreasonable to fear a beating, or worse. If my date was fair-skinned and the police suspected she was White, we anticipated the harassment—and we were never disappointed.

That day when I was singled out of an all-but-me White crowd, I was suddenly ready to die. Somehow, it was no longer worth it. If this was all there was, to hell with it. The cop's face reddened as I told him that and added that he could go to hell, where he was sure to meet at least one of his parents. Pulling me by the arm, he ordered me to his car, sensing my at-the-end-of-my-rope attitude. However, once in the privacy of his car, he abandoned his macho demeanor and tried another tactic. He wanted to be my liberal friend, telling me of the Negro leader he had invited to dinner. I told him that the Negro leader should improve the quality of his White friends.

Exasperated, he reverted back to form: "Shut up or I'll arrest you."

"You can kill me," I blurted out, probably because it was my biggest fear. "Go ahead! Or take me to jail and beat me. Because I'm not going to sit in this car for another minute with your racist ass!"

It was unheard of for a Black person in Detroit in 1963

to talk to a policeman that way. But I was not being brave. I no longer cared. The barrier of death had fallen. Living had taken on a new meaning. Almost trembling, this big, muscular man with his big gun no longer felt that he was in control. In fact, he was not. I could sense his fear that he had lost dominance of my soul. In a mild voice, and with a hard stare, tensely, he told me to leave.

This was a sea change not only for me and that cop, but for the way Blacks and Whites related to one another. The march that I coordinated took place a few weeks later, and 500,000 Black people, virtually no Whites, turned out to follow Martin Luther King down Woodward Avenue, in a demonstration opposed by the Detroit branch of the assimilationist NAACP. Standing beside James Del Rio, the march's marshal, I heard King deliver his "I Have a Dream" speech for the first time. We were all born again that day as a new people.

I have done well in my career, yet I know the feelings and psychological agony of being Black in a White world. For the Black and powerless, my confrontation with the Detroit cop was a relatively mild occurrence. I can understand why Louis Farrakhan makes sense to many Blacks who are trapped and left behind. He tells them that he visits spaceships and that Elijah Muhammad is not dead and that the White man is a devil and that Jews ran the slavery system. There is no proof, but they do not need facts. They are being guided by the radar of pure emotions. The words in his speeches are just props for the anger they feel that has to go somewhere for therapy—rage must be managed, somehow. Or you crack. If you are not Black, you might easily conclude that they are ignorant and backward. They are not. They are human and they hurt. Their very souls ache with the pain of insult and rejection. Only another Black can know this. Most Whites could not psychologically handle being Black— even being wealthy and Black.

We are all in the same boat now, and we will all be thrown into the water together if our economy is bankrupted. An economic breakdown will also mean a breakdown in the social contract. Entitlement-fat Americans are bankrupting themselves with a corrupted form of transfer socialism. In present-day on-the-dole America, sacrificing and changing to solve the oppressive debt crisis appear as likely as Blacks and Whites accepting a common destiny. The committed, courageous leadership the country needs would force the American people to acknowledge responsibility by stopping the handouts and the free lunches. The public should be forced to accept responsibility for its own destructive behavior rather than hide behind demagogic politicians who willingly accept the blame in exchange for remaining in office as professional scapegoats. Ultimately, as creeping socialism corrupts our democracy, the politicians and government will have absolute control over the lives of the voters.

It is worth noting that National Socialism was the ideal of Adolf Hitler and the National Socialist Party of Germany, and that Scientific Socialism was the ideology of the Union of Soviet Socialist Republics. Both regimes were coercive and destructive. Both systems collapsed under the weight of their own oppressive regimes.

Socialism is a failure also in the United States. We have only to look at the "leadership" in the Black Unaccountable Machine (BUM) to understand that.

CHAPTER 3

THE FAILURE OF
BLACK
AMERICA'S LEADERS

The day we see the truth and cease to speak is the day we
begin to die.
—MARTIN LUTHER KING, JR.

America's Black leadership sees its primary function as blaming Whites for the dire problems of the Black community and demanding more government intervention as the sole solution to this predicament.

Even when the government responds with more opportunity—education, welfare, preference programs for the middle class—these leaders are too shortsighted to mobilize the public largesse into a private self-help initiative to attack the collective problems of the Black community. No amount of government intervention can solve the problem of an inept, failed leadership. Therefore, White racism has had a partner, a co-conspirator, in the marginalization of Black Americans. Black economic development and social equality have, in part, been subverted by the very Black leaders who pointed their fingers at Whites.

The combination of irresponsible Black leadership in tandem with White racism and socialistic intervention by government goes to the heart of Black failure in America. Most Whites are thoroughly convinced that today's dysfunctional Black community is evidence of a lack of ambition. Many Blacks, on the other hand, cite slavery and its residual effects as the primary reasons for Blacks' failure to achieve social, economic, and political equality with Whites.

As you may know by now, I disagree with both viewpoints. While America has been and continues to be unfair to its Black citizens, the opportunities provided by the system outweigh this liability. More than a Black point of view, mine is a generational point of view. My generation was taught that educational achievement, moral virtues, and economic organization were far more important to our development than how Whites treated us or how they felt toward us. The feeling among more recent generations of Blacks seems to be that they cannot succeed without the permission of an almighty White population.

I hold that our failed Black leaders and their misguided White liberal "allies" are responsible for the sad fact that Blacks are the only ethnic or racial group in this country that has never become a serious player in the economy. That is why you find Korean-Americans, Arab-Americans, Chinese-Americans, East Indian–Americans, and every kind of newly arrived legal or illegal immigrant controlling the commerce in Black neighborhoods. Black money flows out of Black communities like blood from a wound instead of flowing through it, giving strength and nourishment.

The United States may not be a color-blind society, but the only color of freedom is green here and everywhere else in the world. Everyone seems to understand that but socialists and Black leaders, many of whom are the same, who believe that living off three cents from every dollar spent is a "civil rights" victory.

The failure of Blacks to harness economic power has resulted in the great social and economic divide that still separates their community from the rest of the nation. And the responsiblity for our lack of effective leadership lies with those privileged Blacks who should be the greatest advocates of economic growth. Too often, our best and our brightest have let us down.

Desegregation, Yes; Integration, No

In the campaign for civil rights, the Black rank and file fought for desegregation, an end to state-sponsored segregation, and economic and political equality. But the liberals and their sycophants in the Black protest movement (mostly the lawyers), seduced them with integration, a social elixir that came to be legally interpreted as cultural *assimilation* into the White community. The result was that Blacks who wanted to succeed were compelled to deny their own cultural heritage, to become as White as possible in order to enter the marketplace. It failed because Blacks cannot be Whites, culturally or physically.

Integration is a novel form of liberal racism. It works in insidious and pervasive ways to repress Blacks and keep them marginalized and dependent. Integration discourages capitalistic entrepreneurial thinking. It tarnishes Black learning institutions. It separates upper-strata Blacks from the rank-and-file Black community. It impedes upward mobility by removing role models from the community and eliminating networking opportunities. Integration undermines Black advocates of empowerment and marginalizes them as "nationalist" or segregationist (euphemisms for anti-White). It forces White leadership upon the Black masses through sycophant "Black leaders." In short, integration undermines and stunts development in the Black community.

Black leaders first failed Black Americans when they embraced integration rather than economic equality as the solution to inequality. It was as though they thought just being around Whites would win Blacks acceptance and equality. They don't get it. Black leaders have never grasped the fact that Black people don't have to be "integrated" with Whites to survive. We don't have to go to their schools. We don't have to live in their neighborhoods.

But we do have to come together to save this country. Our mutual survival depends on it. With our workforce becoming increasingly multicultural, we are faced with the necessity of making all opportunities available to all Americans on the basis of character and skill alone. This is crucial not just for Blacks but also for our nation's industrial salvation.

Like most Blacks, then, I support desegregation, but I see integration as a mechanism to marginalize Blacks by preventing their empowerment. Integration, as preached by racist White liberals and socialistic Black leaders, has guaranteed the economic irrelevance of the Black community and led to its political exploitation and social breakdown.

Liberal Racism

I call many of the White liberal allies of the Black community "racists" because they have demonstrated their belief in the inferiority of Blacks by enacting policies that place Blacks at the mercy of welfare and other socialistic entitlement programs for the middle class as well as the poor. This perpetuates dependence by Blacks on them, rather than encouraging Black self-reliance and economic independence. Our leaders, Black and White, have shunned programs that might make the Black masses competitive with Whites.

Many of the problems that Blacks face today—material poverty, dysfunctional families, illiteracy, and so on—can be traced back directly to the organization of the Black protest movement by Whites, many of whom were socialists. That White leadership instilled socialism and its "Where's mine?" mind-set in Blacks, rather than capitalism and its "I'll get my own" attitude.

Although liberal Black Democrats do not describe themselves as socialists and most of them do not consider themselves to be socialists, many are basically socialist in their ideological approach. At the same time, there are Black conservatives who leap at the opportunity to serve the Black masses on a platter to the White right establishment. Tragically, some in the Talented Tenth offer themselves to either wing of the White establishment as their pawns rather than representing the interests of the Black community.

In the current bickering process, the Talented Tenth—
the historical term for America's Black professionals and
leaders—kill each other off with slander on behalf of either
the White left or the White right. I contend that the White
establishment's plantation system controls the Black com-
munity—politically, economically, and socially—through a
slavish cabal of acolytes.

An accountable Black leadership would never subordi-
nate itself. It would be committed to the advancement of its
own people. Black leaders are not accountable to Black peo-
ple, in part because the Black community is so fragmented.
The Black elite don't really need the Black rank and file, and
so they are not responsive to it.

White people rarely understand the inner workings of
the Black community. Often, they have the naive, and racist,
perception that Blacks are politically and socially homoge-
neous, which explains why Whites and the White-dominated
media have difficulty interpreting and reporting on varying
points of view within the Black community.

Not all Blacks are Democrats. We do not all consider
Jesse Jackson our great leader. We are not all liberals. We
are not all poor and uneducated. Some of us can't even sing
or dance or play basketball. There is great diversity within
Black America, and to help make that point, I have devised
the following list of our "tribes."

In the contemporary Black community there are four
primary cultural tribes. Although as a whole the Black com-
munity constitutes a socialist Democratic team politically,
the tribes are as culturally diverse as the general population.
And politically, motivated by inbred values. Overall, group
distinctions have been defined by the formation of the Black
leadership and the class structure of American society with

Blacks on the bottom. This American version of apartheid has adopted the former South African structure as a prototype: Afrikaners (Whites) at the top of the socioeconomic pyramid, followed by the coloreds as a buffer group (a standard ploy of colonizers) and the most severely oppressed group on the bottom rung of the social order. In the United States, a Black professional class—today's Talented Tenth—serves unwittingly as a "colored" buffer group in the traditional divide-and-conquer strategy of dominant groups.

The Tribes of Black America

1. *The Talented Tenth*. This is the most influential Black professional and intellectual class. The model for this group in terms of educational achievement and wealth was the original Talented Tenth, an upper crust of freeborn or newly freed "coloreds"—most of whom were racially mixed. In the 1920s, they numbered about ten thousand, or one in one thousand of the Black population. The concept of an elitist Black oligarchy was originated by intellectual assimilationist W.E.B. Du Bois at the beginning of this century. Unfortunately, as a result of their inbred elitism, many in today's Black leadership confuse affirmative action with a meaningful program for the Black masses.

Black elitism has never had much faith in the Black rank and file. Du Bois, the Talented Tenth's paterfamilias, made that plain: "The Negro race, like all races, is going to be saved by its exceptional men." Few Whites comprehend class distinctions among Blacks, but affirmative action is viewed by the Talented Tenth as a program benefit to the upper classes. When a Black millionaire can qualify for a program designed to help the economically disadvantaged, it is not affirmative action, it is upper-crust welfare. This distinction has escaped the Talented Tenth.

Some of the Talented Tenth are very clear about their anti-Black identity. Often, they flaunt their racial self-loathing. Comedian and actress Whoopi Goldberg, who specializes in "nigger" humor, is among the most extreme group of honorary Whites. Others include television talk show host Bryant Gumbel and an obscure New York cartoon character, Michael Meyers, who has been discovered by the New York media as a Farrakhan-NAACP basher and a "civil rights" leader. They are all honest and openly proud of being honorary White people. As Meyers said, there is no "such thing as Black culture. . . . I don't know what it means to be proud that you're Black."[1]

2. *The Black Unaccountable Machine (BUM).* Just as the early assimilationist NAACP's original mission was literally for the advancement of *certain* colored people, the literal function of today's Black leadership is to serve as a shield for White liberalism and its institutions. It is a bait-and-switch operation. The Black Unaccountable Machine's leaders legitimately decry the oppression of the Black masses, but instead of doing something to alleviate that oppression, they endeavor mostly to enhance opportunities for the upper classes and liberal racists.

The beneficiaries of boycotts and Democratic Party voter registration drives are the Black elite, not the Black masses. Boycotts of corporate businesses require White firms that are often arbitrarily charged with racism to cut business deals with Black-owned firms that often give financial support to the sponsors of the boycotts. Sometimes these boycotts are turned into payoffs for elitist members and supporters of a self-imposed Black government that is not accountable to the people it purports to represent; therefore the name: BUM.

Although the BUM enjoys media dominance over Black thought, it is a moral and intellectual disability. It is hard to explain in rational terms how a community that has equality as its goal clings to the tradition of a group of unaccountable and dictatorial grandees.

The pretext for this oppressive conformity is group solidarity against racism. However, if an ethnic or racial community is to progress, its shared beliefs must be based on a truth that makes it competitive. What was in part motivational symbolism during the civil rights era of integration is retrograde dependence today and contributes to a permanently noncompetitive Black underclass. That is why unity based on uniformity is destructive; it disconnects from reality. In the real world, this culture of Black-on-Black oppression and authoritarian socialism has created a world of losers. It perversely undermines responsible private behavior and fosters an attempt to live life backward. The BUM claims to guarantee success without controlling the means of providing it.

The first step toward success is realizing who you really are, then performing the tasks necessary to get what you want. What is missing from the Black experience is the use of heritage as a source of pride and self-esteem, which provides the basis of productivity. The Black community is not competitive because it is not productive. And it is not productive because it does not value itself. As a result, Blacks live the American Bad Dream.

The most insidious factor in the demise of the Black community is the historical evolution of Black leadership's interests from service to self-service. For instance, the Blacks in the federal legislature were representatives of a Black liberation movement when they were first elected to office by a community-oriented population in the 1960s. When the Congressional Black Caucus was formed in 1970, its purpose was to further a Black agenda of individual and group responsibility to foster community growth and self-empowerment. During those days, the CBC's members fought both White Democrats and Republicans who resisted the aspirations of the Black community in its quest for self-determination.

Over time, however—as it came under the influence of integrationist party regulars and was bought off by PACs, liberal racists, and corporate lobbyists—the Congressional Black Caucus of 1970 evolved into the Democratic Socialist Congressional Black Caucus of today. It is now largely an adjunct of the Democratic Party, the labor unions, and White carpetbaggers. Its primary role in the Black community is as the chief affirmative action benefactor for the well-off Black elites.

The annual upper-class CBC soiree alone generates an estimated $500 million (one-half billion dollars) for direct and related expenses to attend a week of galas in non-Black-owned hotels and facilities.[2] This is the current status of what was once a proud movement to elect to office Blacks who would facilitate the growth of those at the bottom in the Black community. The most lamentable traitors among Blacks are not the greedy and immoral leaders, but those liberal and conservative Talented Tenthers who aid and abet the destruction of their own community because they are too cowardly to oppose the BUM.

The Talented Tenth Colored Oligarchy is a subdivision of the BUM. It has two primary branches.

A. Black Leaders. Some widely recognized "Black leaders" are not leaders of Black Americans in reality. They are media creations or self-promoters. And their false leadership distracts from the fact that Blacks are a complex, multifaceted group.

The media are too lazy, or too ignorant, to realize that not all Blacks are liberals. Media reports, for example, ignore "Republican Black leaders." Instead, the liberal media intentionally label Black Republicans as "conservatives" because they are aware that the term carries a negative connotation among Blacks.

As patriarch of the Talented Tenth, W.E.B. "Willie" Du Bois, was a brilliant, confused, elitist, arrogant, socialist, Communist iconoclast whose single greatest achievement, in my mind, was to quit the assimilationist NAACP in 1934 over its obsessive integrationist policy. His modern-day minions—integrationists and socialists (called "progressives")—maintain a symbolic image as freedom fighters. They would more aptly be called "plantation warriors," because even after all these years, they are still mere intermediaries in a plantation system of external White control.

Renaissance writer Claude McKay described the socialists and Marxists of this cabal in his 1940 book *Harlem: Negro Metropolis* as "the black butt of Communism."[3]

B. The Great Society Hustlers. This is another branch of the BUM, made up of an emerging class of wannabe "Black leaders" who are acolytes of the Talented Tenth elite but are not ranking members because of their cultural traits and lack of formal education.

3. The Black Masses. Many in this rank-and-file category are marginally middle-class, others are poor and working poor, and about two million are in the underclass. What they have in common is that they are exploited by their own Black leaders. In my opinion, Black writer Claude McKay made the most insightful observation of the class relationships in the Black community. It is as accurate today as it was when it was written in the 1920s:

"Perhaps there is nothing more significant in the social history of the United States than the spectacle of the common Black folk in overalls and sweaters agitating and parading for jobs for apathetic white-collar Negroes."[4]

Most ordinary Black people rarely question the legitimacy of their Black leaders and their White liberal manipulators. Many Blacks hold to the myth that Whites, and particularly White Republicans, are the primary obstacles to Black empowerment. And so the Black elite maintains its position and control over the masses by failing them and blaming the failure on Whites, and specifically on Republicans, some of whom deserve the condemnation. But even if the White conservatives fell in love with Blacks and the White liberals stopped exploiting them, would anything change? No, because the Black liberal elite and its wealthy rip-off artists would still be there to bleed the trusting Black masses.

4. The John the Baptist Brigade. This fourth tribe is perhaps the least visible and is all but unknown in the White community. It works mostly behind the scenes, with the exception of a few warriors who openly battle with the Talented Tenth elitists and the liberal White racists. I consider myself one of the more confrontational members of this tribe. We come from all three of the aforementioned groups: the Talented Tenth and the Black Masses and even a handful who have escaped the graft of the BUM crowd and fled the plantation. We are modern-day Maroons, the revolutionaries during slavery who escaped the slave plantation only to return to attack it again and again, leading off new recruits to freedom each time.

I call these Blacks the John the Baptist Brigade and regard them as an apostasy because we are forever crying out in the wilderness—knowing that we will be ridiculed and marginalized by the very people we love and by the venomous Black intolerants who allow no freedom of speech from other Blacks. In one instance, in a puerile defense of quotas for incompetent Blacks, *USA Today*'s DeWayne Wickham, the irascible Black columnist who has never seen a liberal Democrat he didn't admire, accused the head of the National Urban League of "conflicted behavior" (consorting with the enemy) because he believes that people can oppose affirmative action on a basis other than racism. For that, Wickham imposed the "black tax" and accused the venerable Urban League of "mealy-mouth, fence straddling ways."[5] We also recognize that, like John the Baptist's, our heads will be (figuratively) cut off by an outraged status quo of Black vested interests and liberal racists in the media. Perhaps, if we're successful in waking our people up to their God-given innate talents, we'll be crucified as was Jesus for bringing truth to His people and the world. My belief is that everyone should seek his or her own crucifixion. To be nailed to the cross of truth is God's eternal blessing.

Many of us are loyal to the struggle of our people but not to the political gangsters, Black and White, who have hijacked it. Other members of this Black truth brigade include Imam Wallace D. Muhammad, spiritual leader of one million Americanized Africans who practice orthodox Islam; Eugene O. Jackson, CEO of World African Network; Pulitzer-prize-winning columnist William Raspberry of the *Washington Post*; Preston Wilcox, an educator devoted to Harlem's Blacks; Robert Woodson of the National Center for Neighborhood Enterprise; Elizabeth Wright, publisher and editor of her own newsletter, *Issues & Views*; and Bernard Kinsey, former cochairman of Rebuild L.A., a commission appointed after the riots following the Rodney King verdict. There are scores of others whose goal is for Black Americans to control their own destinies by taking responsibility for our social and economic development.

The Political Failure

Sadly, I have concluded that most of the Black population is virtually enslaved by an entrenched Black leadership manipulated by White liberals. This remains the case now just as it was almost a century ago. In his later years, when he realized that he had been used by Whites, Du Bois himself predicted this would happen. One of many of Du Bois's late-blooming intellectual insights into racism appeared in *The Crisis* when he wrote to expose the insidious liberal-White agenda that he had previously promoted and noted that the Black "has nothing but 'friends' and may God deliver him from most of them, for they are like to lynch his soul."

Just as the Jews overcame poverty and anti-Semitism, the Black masses could have succeeded economically and educationally in overcoming marginalization created by racism—despite the opposition of White racists—if a different and more visionary leadership had been in place. Apparently, neither the Black Talented Tenth nor their White socialist-liberal mentors believed the Black masses were capable of getting the job done. And neither does today's Black leadership, the BUM.

The Niagara Movement for Civil Rights began in 1905 to "abolish all racial distinctions in the United States." It was self-defeated and ended by 1909, not because of White opposition, but because of Black ineptitude. The leader of that campaign was W.E.B. Du Bois.

Du Bois was aligned with White socialists who organized Talented Tenth intellectuals as their followers. This was functionally the "assimilationist" NAACP, not to be confused with the modern-day "empowerment" NAACP led by Black capitalists who believe in free-market solutions to the Black predicament.

The liberal Whites who controlled the early NAACP and their handpicked Black leaders shunned free-market competition with other ethnic and racial groups and instead advocated dependence on the largesse of government and White people. This socialistic and assimilationist philosophy has influenced everything from legislation to literature. And it has been corrupt from its earliest days, as I will show in the next chapter.

The Black Virus of Socialism

Although history has not followed Karl Marx's projections, many Black scholars continue to abide by his failed philosophy and indoctrinate the young with a duplicitous brand of

Marxian socialism that has succeeded only in blunting the competitiveness of the Black community.

One of the fellow travelers, Manning Marable, a Columbia University professor and vice chairperson of the Democratic Socialists of America, who comes from a family of Black entrepreneurs, not only opposes the pursuit of profits but indicts all Black entrepreneurs as well. He writes that these "petty capitalists" are only interested in their own selfish end of making money, and by "exhorting Black consumers via Black nationalist appeals to 'buy Black,' " they exploit their own people. These "petty capitalists" also include any Black who works in finance or at a bank or who, like him, works for a "white-owned corporation." All of them will do just about anything to make a profit, Marable writes in his socialist manifesto. Money is also channeled from "the white private sector" to Black firms "through NAACP-style pressures."[6] In *U.S. News & World Report,* an obscure Marable named himself as head of one of three groups vying for leadership of the Black community.[7]

In his quest to impose Socialist-Marxist-Communist doctrine on Black thought, Marable's propaganda has sullied the reputations of some of the Black community's real giants, but it is all part of his avowed campaign to "undermine" the institutions in the Black community by any means necessary. Nathan Wright, organizer of the first Black Power conference, has been indicted by Marable as a "Reagan apologist" for believing that socialism and Marxism are failures—an opinion now shared by most of the rest of the world. Percy Sutton, Malcolm X's lawyer, a staunch Democrat, founder of a Black broadcasting company and past Manhattan Borough president of New York, is also attacked by Marable. Marable became livid when Sutton called for more police for the crime-riddled inner cities. That is a neo-conservative act, according to the radical socialist. Another Marable target has been John Johnson, the founder of *Ebony*, *Jet*, and many other enterprises. Johnson, who provides jobs for thousands, committed the socialist crime of making a profit, Marable said.

Marable's philosophical comrade, Marxist-oriented Ben Chavis, severed head of the NAACP, also condemns profit making for Black people. His soulful socialist explanation says that "profit-oriented capitalism is a way of ordering life fundamentally alien to human value in general and to Black humanity in particular."[8]

In effect, these socialists and Marxists are still restricting their own people's development nearly a hundred years after their Socialist-Marxist predecessors aided the re-enslavement of Black people through assimilation and the destruction of economic organization in the Black community. And like the servant with one talent, they have been blinded by their true religion of socialism to the fact that the talents (self-help) of their people are indispensable to their survival and progress—which is God's will.

Most Blacks, unfamiliar with the historical legacy and failures of Communism and Marxian socialism (from which these grandees take their cues), miss the significance of these bourgeois ivory-tower campus intellectuals. But the paterfamilias of the Black Marxist rat pack is Harvard's (and formerly Princeton's) Cornel West. The wannabees, like Manning Marable, are a little rough around the edges, don't have the White media connections, and earn only around $200,000 a year, including speeches and books—a slow year for West. At the White capitalist academy, they are serious about polluting young minds with a phantom socialist religion while living high on the hog of "capitalist exploitation."

If ever there was a petite bourgeoisie, it is these rich oracles of progressivism.

These "educators" are quintessential examples of why Blacks have failed to produce a viable leadership from those who pass through their classes. Their students are taught that capitalism is responsible for racism and they become anticapitalists. Because West graduated from the White academy (Harvard), someone has dubbed him "the preeminent African American intellectual of his generation." Paternalism ad nauseum. "West's published work is an endless exercise in misplaced Marxism,"[9] wrote the literary editor of *The New Republic*, Leon Wieseltier.

Why do so many Blacks in higher education become Marxists? The benefits package, according to West, is exhilarating: "The Marxist model . . . provides entry into the least xenophobic white intellectual subculture available to black intellectuals."[10] In addition to social intercourse with the best White people, it also satisfies the need of the Marxist bourgeoisie intellectuals for "recognition, status, power, and often wealth,"[11] West writes.

Harvard reportedly pays West $150,000 a year. And then there are honoraria and book royalties, say another $500,000. Not bad for someone who teaches his students that capitalism doesn't work. In addition, the media, as West puts it, aggressively promotes "the prophetic role" (his own self-endearing term) of Black Marxists. "Marxism provides . . . highly visible leadership roles,"[12] and he recommends "the Marxist model" to Black students if they are ever to amount to anything as scholars. The payoff for puerile Marxism is White friends, big bucks, your face on TV, and your name in the *New York Times*, where they anoint you the smartest Black in the world.

Standing their own class theory on its head, a few of

these Black Marxists have become the sort of $150,000-a-year elitists they traditionally rail against—the petit bourgeois professor who has chosen a corporate career in the heart of what they refer to as "this corrupt capitalist system." While some of West's fellow anticapitalists sign lucrative book deals, receive substantial honoraria, and employ high-powered White agents, they counsel the young and impressionable under their charge to avoid the evils of free-enterprise, free-market American democratic capitalism. They are true to the tradition of the Talented Tenth upper crust. Afraid, I suppose, that some of the rewards of a capitalist market economy will rub off on other Blacks, they appear to advocate noneconomic socialism for others and wealth creation for themselves. This is typical Marxian duplicity: The ends justify the means if they are to gain "hegemony," as Marable puts it. It is a morally deficient double standard from people who profess socialism, which in reality is nothing more than a quasi-religious outlook masquerading as economic theory. In fact, the only historically predictable features of Communist systems are elites who dominate as a privileged ruling class and the prevalence of totalitarianism and mass murder. The socialist utopia they promise has never existed anywhere on earth.

Du Bois was the quintessential Talented Tenther Communist. Understanding him philosophically is the key to understanding the mind-set of many contemporary Black intellectuals and leaders. Moreover, in Du Bois one can identify the contradictions and conflicts that became institutionalized in Black leadership dogma. He was the main Black co-conspirator in the corruption of Black leaders by White assimilationists and socialists when they organized the Black protest movement.

Harold Cruse, author of *The Crisis of the Negro Intellectual* and *Plural but Equal*, writes about the White liberal's legacy of "noneconomic liberalism," whose vestigial form could very easily be called socialism. This virus of socialism entered the body politic of Blacks at the turn of the century and has since resulted in a major epidemic of economic and social pathology. Cruse explains the foundation philosophy for the twentieth-century reenslavement of Blacks in his highly acclaimed book *Plural but Equal.*[13] Noneconomic socialism was the preeminent article of faith when the White-led Black protest groups arbitrarily took command of the Black community.

The Last Great Leader

The most notable exception to this hypocrisy among Black leaders, many of whom are holding guaranteed reservations on the lowest rings of Dante's hell, was Martin Luther King, Jr., who understood the need for economic self-sufficiency as well as an inclusive desegregation philosophy that he misclassified, as most people do, as "integration." King was unique among integrationists when, at the height of the civil rights movement, he espoused economic self-sufficiency as well as pluralism.

For the most part, the integrationist Black leadership, with the exception of self-help advocates and entrepreneurs, has always been so desperate for White approval that it has historically opposed any demonstrable equality of its own people.

For instance, as Elizabeth Wright explained in her *Issues & Views* newsletter, at the dawning of this century

Chicago's Black elites crusaded against Black-owned institutions because they "feared that if blacks appeared capable of too much self-sufficiency, whites might come to look upon them as not desirous of integration." These integrationtists went so far as to undermine self-sufficiency projects that they feared would send a signal that Blacks were capable of supporting themselves. Out of a fear of being themselves and making their innate equality manifest, the Black upper crust crushed any expression of self-reliance.

When the Black icon and journalist Ida B. Wells tried to start a kindergarten for Black children in Chicago, she was thwarted by the Black leaders who preferred to wait and see if Whites would subsequently allow Black children to attend a White-run kindergarten. Wright explains that a group of Blacks were defeated by Black leaders for the same reason when they tried to start a YMCA for Black youths. Black integrationists believe that Black culture must be denied and self-sufficiency deferred at all costs, if one-way integration with Whites is ever to be realized. Implicit in this assimilationist paradigm is the essential inferiority of Black culture and the condition that Blacks not be intellectually, socially, or economically competitive with Whites. This is also the guiding tenet for busing Black children to White neighborhoods.

The Missed Path to Black Empowerment

The original sin of Black leaders was to accept as fact the myth that White people would or could solve the crisis of Black people. Instead of attacking material poverty with education and economic self-determination for the masses, liberal Whites and their socialistic Black minions emphasized dependence on legislation and government intervention. Even to this day, when their legislative protests are successful—as in the case of affirmative action—it benefits primarily the Black elite and not those Blacks who truly need assistance so that they can compete in the marketplace.

In order to curry favor with the powerful White liberal establishment, the wealthy German Jews—the early immigrant merchants and bankers and garment makers who had attained economic power and status—parented the Talented Tenth Black leadership as a guardian does an orphan. The Jewish establishment helped Blacks, but not in the same manner that it assisted another poor immigrant group, the second-wave Jewish immigrants from Russia and Poland.

Although they arrived at the same time and shared many of the same social and economic problems, the poor Black immigrants and the poor Jewish immigrants have fared far differently. The Jews have generally succeeded while the Blacks have not. This is because upon their arrival over a hundred years ago, these two groups took different paths. And they did so at the direction of the liberal Gentile and Jewish establishment.

The Russian and Polish Jews were encouraged by their wealthy Jewish patrons to develop their own economic infrastructure and to compete in the marketplace. But the same

Jewish establishment did not advise the Black masses to become economically competitive. Instead, they fostered among them a sense of dependence and reliance on entitlements and paternalism from Whites. They did this by encouraging Du Bois's idea of the Talented Tenth elite, which would, with White sponsorship, take care of the needs of the Black masses.

But, for the most part, the Black elite, which has sought assimilation with Whites rather than economic competition with them, has taken care of only itself. The Black underclass is partly the result of this abandonment.

What if history had written a different scenario for Black leadership—one in which the early Black rank-and-file followed self-help advocates Booker T. Washington and Marcus Garvey instead of the Talented Tenth elitists who shunned and exploited the masses? What if Washington or Garvey had allied with the wealthy uptown Jews and organized a Black mass movement built upon legislative protest and commerce—butcher shops, grocery stores, and other middleman businesses—along with the values of God, education, savings, and the work ethic?

If they had developed self-reliance and economic strength, Blacks would have been elevated in American society just as the poor Russian and Polish downtown Jews and other immigrant groups were. If Blacks and their elitist Jewish allies had adopted that strategy, their communities might be united today as partners in a common commercial market. But it did not happen, and now both groups are reduced to a symbolic Black-Jewish "special friendship" that is built not on mutual strength but on Jewish paternalism.

The Black elitists' identification with Whites explains why Talented Tenth Blacks created a culture in which lighter-skinned Blacks were favored over darker Blacks—a prejudice that still lingers in many sectors of the Black community. This prejudice was so entrenched among the Black elitists that the dark-skinned Garvey was nearly thrown out of the Detroit church of the Rev. Robert Bagnall, an early NAACP director. It seems Garvey made the mistake of sitting up front with the light-skinned Blacks.

This amazing cult of racial self-denial also explains why Dubois era Talented Tenth Blacks mimicked the lifestyles of the Gentile upper crust even though most lacked the resources to bankroll such a standard of living. Elitist Blacks yearned to be like their Gentile and Jewish idols and often expressed frustration and contempt for any reminders of their Blackness. Witness the Black child raised as Willie who reinvented himself as W.E.B. Du Bois, the Black man who whined incessantly of his double consciousness—his "twoness." Of course, an identity crisis is inevitable for those Blacks who attempt to assimilate as Whites.

The Sin of Dependence

Have the Talented Tenthers and their modern-day incarnations—Black socialists, progressives, integrationists, and ivory-tower campus intellectuals—done any good? *Yes.* I won't deny it. Even some great things have been accomplished. They exhausted the Fourteenth Amendment, primarily with legislation that struck down state-sponsored segregation, and scored other momentous legal victories over discrimination. There were also political achievements, and in the last few decades a record number of Blacks have been elected to public office. Just about all, unfortunately, are in the service of the Democratic Party and not the Black community.

Our Black leaders have always been, and remain, talented men and women, though most of them have never stepped back and considered the destructive results of their actions as agents controlled by outsiders with suspect agendas. They have done little to truly solve the real problems of the Black masses, yet most of these anointed Black leaders have personally benefited from the status quo by garnering honors, recognition, appointments, prestige, and support from the White Establishment. As a result, Black leadership has been compromised and uniquely marginalized.

Although they cannot be blamed entirely for their failures and inadvertent betrayal, Black leaders should not be lauded for them either. They did not deliver the masses out of ignorance and poverty as they promised, and they did not lay an economic and educational foundation for growth. They did not even try. Furthermore, they sometimes intentionally destroyed those who tried to build one—and they continue to do so today.

For the Talented Tenth, Black opportunity has always meant advancement for the elite only. The primary vehicle for Black opportunity, in the elitist vision, are preference programs for a privileged class of Blacks—professional jobs through middle-class affirmative action; business set-asides with no accountability to hire poor Blacks or support institutions in the Black community; and quotas without a quota on the quotas. These are integrationist monuments to self-hatred, and they give official sanction to the myth of Black racial inferiority.

Black Brownshirts

Some of the members of the Talented Tenth do recognize that the elevation of the poor and the training of the unskilled are imperative. They want to shift the Black agenda toward self-determination. But enormous social, political, and economic pressures are exerted on any member of the Black elite who breaks rank. Those who criticize liberal socialism face loss of jobs and income, media attacks, and pariah status within their community.

There is little room in the Black community for freedom of expression, or for many of the other freedoms that Blacks demand from Whites. The Black community is run by an oligarchy of brownshirt plantation overseers who could improve their Nazi impersonations only if they spoke German. The Black brownshirts move in swiftly when any of the rank and file publicly stray. But it is in the defense of the Democratic Party that they are at their most vicious.

Most members of today's Talented Tenth believe that their contributions to the civic and social uplift of their community is in the best interest of the poor and uneducated. Moreover, they do not psychologically understand their role as a ruling-class buffer for liberal White power. Clueless, they genuinely believe that affirmative action, which elevated privileged people such as themselves, is doing the same for their entire community. Often, these are good, honest people with integrity—among the best people in the American society. They are ministers, scientists, teachers, doctors, engineers, lawyers, corporate executives, entrepreneurs, social workers, administrators—they include even most Black politicians. And although they ridicule their own elitist leadership in private, they will attack anyone, Black

71

or otherwise, who dares to say the same things in public. Unfortunately, this misplaced loyalty to an ineffective leadership perpetuates racism more than it helps the poor and disadvantaged.

Blazing Trails

Blacks are not going to win political power in a sideshow with maniacal ringmasters for leaders. The Black community cannot move forward without the resources controlled by today's Talented Tenth leadership, but in spite of its human and financial wealth, the Black upper crust doesn't have its political sea legs or even much insight into how it has become trapped in a time warp. Today's Talented Tenth is incapable of helping the Blacks who need help most because of the class-bound political naiveté and poor grasp of its own elitist history. It is a legacy of that oligarchy's narrow philosophy, exploited by contemporary opportunistic leaders, that has placed today's Talented Tenth leaders in the tragic position of abetting the destruction of their own racial community.

Most Blacks that I know are just like the average Red, White, Yellow, or Brown person. And like the average person, they depend on their leaders to blaze the paths for them. If the leaders are not good trailblazers, the lives of the followers are diminished. In the American capitalistic free-market economy where the majority of leaders of other racial and ethnic groups are blazing trails, many Black leaders are passive participants. Many even resent free-market opportunities.

Instead of playing the game of competitive economics,

Black leaders either plead the case of militant socialism or plead for government handouts. Most of them have no ideological stake in the semantics of socialism, in which government controls the means of production and the allocation of resources. Very few will admit to being practicing socialists, but the Black community's leaders are essentially socialistic in practice—even if they disavow it philosophically. And that makes it even more difficult for those who have put their trust in a leadership that insists on further corruption of the market economy with socialistic controls.

CHAPTER 4

CONSPIRACIES AND BLACK AMERICA

He who does not understand the past is doomed to repeat it.
—GEORGE SANTAYANA

It is difficult to prove that an entire nation is bent on your failure, but when the socioeconomic gap between Blacks and the rest of America has persisted for three hundred years, one of two things must be true: Either Blacks are genetically inferior, or something has conspired to keep them from succeeding as a people. I believe the latter is true, and I am not alone.

In a *New York Times* article entitled "Talk of Government Being Out to Get Blacks Falls on More Attentive Ears," reporter Jason DeParle wrote that conspiracy theories are gaining credence because of a growing awareness of an American past featuring real plots against Blacks—ranging from slavery to the FBI's infiltration of civil rights groups.[1] Many Blacks recall the Tuskegee, Alabama, syphilis study in which the U.S. Public Health Service and the Centers for Disease Control and Prevention used Black men as nonconsenting guinea pigs. The mostly White doctors watched as the poor Black men, who thought they were receiving treatment, wasted away over a forty-year period. Three hundred ninety-nine of them died from 1932 to 1972.

In this chapter I will provide some episodes from history that have never appeared in standard textbooks. And I will tell you of some of the conspiracies perpetuated against Blacks by enemies seen and unseen, both from the outside and, tragically, from within.

The Enemy Within

In the spring of 1993, the *Memphis Commercial Appeal* newspaper published a series of stories by reporter Stephen G. Thompkins that made public for the first time the fact that Joel Spingarn, the White board chairman of the NAACP from 1914 to 1919 and its president in 1930, operated a network of Black spies within the Black community under the direction of the U.S. Army.[2]

Thompkins's sixteen months of research on the "largest domestic spy network ever assembled in a free country"[3] turned up documents including memoranda, meeting notes, and diaries from archives and private collections. Some of the documents are still classified, and the pertinent ones were made available to me by Thompkins, including information that was not released in his series.

When America entered World War I in April 1917, one

in ten of its citizens was Black—twelve million people. As America drew closer to war, Army leaders suddenly realized how dangerous Blacks could be to national security if they organized against their government. Blacks were well positioned to thwart the war effort if they were so disposed. There were 27,000 Blacks employed in shipbuilding, 75,000 in coal mines, 150,000 working the railroads, 150,000 in telegraph, postal, and telephone communications, and 350,000 in various industrial plants.

"Negro Unrest"—as it was termed by the U.S. Army's Military Intelligence Division (MID)—was a major fear of President Woodrow Wilson's administration and the War Department. After declaring war on Germany in 1917, the United States government effectively declared war on all Blacks in America as suspected traitors. The MID created an internal security network to spy on every outspoken and prominent member of the Black community and to recruit Blacks to spy on each other. Government intelligence against Blacks continued into the 1960s, perhaps beyond.

When the Black spy network was organized, the President and the government assumed that because Blacks were discriminated against, they would not be loyal to America during times of war. Blacks were considered to be the enemy within. If that were true, Joel Spingarn's case would be the story of the enemy within the enemy within.

Students of Black history generally know just a few things about this early White NAACP leader. The NAACP's highest honor is presented annually in his name. He is also known as a fierce enemy of Booker T. Washington, the Black educator who advocated self-reliance and economic empowerment. Spingarn was an Austrian Jew who played down his heritage, preferring to call himself an "assimilated American." Jewish tradition emphasizes philanthropy and this cultural empathy was observable among most Jews who were allied with the Black protest movement. Spingarn, however, seems to have been torn between this tradition and his enthusiasm for assimilationism.

Few Blacks know that Spingarn ran a spy network against Blacks, a domestic intelligence operation that served as a forerunner of the COINTELPRO FBI-led counterinsurgency program of the 1960s. In the service of this government spying operation, Spingarn used his position at the NAACP to gather critical information on the Black community. Among other covert acts, he turned over to the MID the NAACP membership list of all 117 NAACP branches throughout the country, with the names and home and business addresses of their top officers.

Agent Spingarn, who had taken a leave of absence from the NAACP during this period to serve as a major in the MID, was assigned to both counterintelligence work against "Negro subversion" and also against the International Workers of the World (IWW)—a radical union and forerunner of the AFL-CIO. Throughout his tenure at MID, Spingarn kept in constant contact with NAACP secretary John R. Shillady, a White man who had taken over the day-to-day running of the NAACP. Shillady sent a weekly "situation report" to Spingarn of major events and issues through an anonymous Washington, D.C., post office box.

With Black MID agent Lieutenant T. Montgomery Gregory, Major Spingarn took on the secret assignment of infiltrating and recruiting America's Talented Tenth. Spingarn's target was the Black aristocracy. He set about recruiting people he knew from his NAACP travels. In particular, Spingarn knew the major publishers of Black newspapers around the country—a primary source for spreading his NAACP fund-raising message. All of this was done very quietly.

Spingarn and Gregory spent a lot of time in the Washington-Baltimore area, which was home to almost six hundred prominent Black families whose business interests reached throughout the South. They vacationed together at Harpers Ferry and Saratoga—two places where Spingarn sent MID agents to spy on their comings and goings and their associates. Also under MID surveillance were several hotels frequented by wealthy Blacks at Highland Beach, New Jersey. Spingarn's agents posed as dishwashers at Washington's Mu-So-Lit Club.

His agents infiltrated the exclusive Foster Whist Club in Baltimore. Spingarn also actively worked to infiltrate an organization called the Association of Oldest Inhabitants of Washington, and he had many contacts in the National Negro Business League. Prominent Black spy recruits included Robert Moton, Booker T. Washington's successor as president of the Tuskegee Institute.

It has also been discovered that Spingarn provided the MID with confidential information long before Lieutenant Colonel Ralph Van Deman, head of the Army's Military Intelligence Division, lured him into becoming an official agent on May 18, 1918. In a June 19, 1918, memorandum, Major Spingarn advised Colonel Marlborough Churchill, the commander of military intelligence, under the headline "Intelligence," that he had implemented plans to conduct counterespionage among American Blacks.[4]

While Major Walter H. Loving, one of six Black staff MID spies, organized and ran the Black Preachers Network as a spy unit, Spingarn concentrated on recruiting the Black aristocracy. Many Blacks who were deemed national security threats were handled by Van Deman's agents or Loving's operation. A good example is William Monroe Trotter, the fiery owner of the *Boston Guardian* newspaper, who wrote about Black rights. Arrested on a number of occasions for rioting in Boston, the defiant Trotter was considered an enemy of America by Van Deman.

Spingarn was not the only early NAACP leader to work for Army intelligence. It is generally known that W.E.B. Du Bois actively sought appointment to the U.S. government's intelligence division in 1918, but in his research, Memphis reporter Thompkins uncovered a confidential memo dated June 10, 1918, from Spingarn to Colonel Churchill of the MID. In it, Spingarn wrote that Du Bois, the "editor of chief colored magazine, *The Crisis* (monthly circulation 70,000), . . . has promised (a) to submit all matter in magazine to designated person in advance of publication, and (b) to make his paper an organ of patriotic propaganda hereafter."[5]

It is apparent that Du Bois agreed to let the NAACP's *The Crisis* become the propaganda tool of a special MI-4 (counterespionage) intelligence unit on "Negro subversion." Du Bois, who is the hero of today's Black Marxian socialists and progressives, was a brilliant but racially ambivalent man. From the information gathered by Thompkins, I have to question his commitment to Black people. His agreeing to join Spingarn's spy network against the Black community further suggests that Du Bois was for hire to the highest bidder.

When his duplicity became public, Du Bois was restrained by the NAACP executive board and upbraided by public opinion. In light of the public controversy and internal opposition to Du Bois, MID promptly withdrew its offer.

Nevertheless, Black historian Dr. David Levering Lewis writes in his biography of Du Bois that the "deal implied cold calculation . . . and it was unworthy of Du Bois. . . .'"[6] It was also duplicitous of Du Bois to tell an NAACP official that he was not offered the spy captain's position until June 15, when in fact Du Bois had already written what many characterized as a progovernment editorial in *The Crisis* on June 6.

Du Bois, the intelligence propagandist, lied to hide the fact that he was already working as a spy. The Army offer was made on June 4—two days before he promoted the military's point of view on the prosecution of the war. Years later in writing his autobiography, Du Bois never mentioned "the Military Intelligence venture,"[7] Lewis explains. It seems that being a self-admitted propaganda agent—as he had been for the White-led NAACP all along—had little impact on his conscience. As the NAACP's employee, Lewis writes, Du Bois "saw himself as an intellectual turned propagandist."[8]

Motive, Means, and Opportunity to Kill

By the 1960s, MID had developed into a greatly expanded and sophisticated American intelligence community, and Black America was still under surveillance. According to Thompkins's *Commercial Appeal* article, Army spying on civilians "expanded in the '60s because the FBI and local police forces proved unreliable."[9] In May 1963, military intelligence started using U-2 spy planes to watch Black protests in Birmingham, Alabama. More U-2 sorties on Blacks followed, as did even more expensive surveillance by the more advanced SR-71.[10] But it was the fear of Martin Luther

King, Jr.'s influence on Blacks that moved the U.S. military to prepare for an outbreak of war with Black America in April 1968, the month that King was assassinated.

When riots broke out in Detroit in the summer of 1967, of the 496 Black men arrested for firing rifles and shotguns, 363 told disguised Army intelligence agents that King was their "favorite Negro leader."[11] This alarmed a military establishment with insufficient war power within the country to halt an armed rebellion led by the charismatic Black leader. The military leaders felt King was taking advantage of them because of the Vietnam unrest and the military's limited ability to respond to violence at home.

When it came to King, these "West Point geniuses" had no "clue" as to how to stop him, the *Commercial Appeal* reporter wrote.[12] If they did have a plan, there was still the problem of a domestic troop shortage because of the Vietnam War. If the expected April uprising occurred in the United States, there were not "enough combat-type troops to react,"[13] Ralph M. Stein, analyst in the Pentagon's counterintelligence bureau in 1968, told Thompkins.

Their intelligence-gathering convinced them of "King's plans to ignite violence and mayhem" nationwide in April 1968 with his Poor People's campaign in Washington. Their repsonse was to mobilize the United States military into a state of preparedness against the Black community. Military planners "covertly dispatched Green Beret teams to make street maps, identify landing zones for riot troops and scout sniper sites in 39 racially explosive cities. . . ."[14] Some 124 cities were scouted as war zones, and troublemakers were identified.

During the Kennedy and Johnson years, an additional four hundred Blacks were added to the FBI's "Security Index" of ten thousand "racial agitators" listed by their "degree of dangerousness,"[15] reported Kenneth O'Reilly in his book *Racial Matters*. The listing was begun in 1939, undoubtedly by the Military Intelligence Division, because the index was formally called the "Bureau War Plans." It operated on the same premise as the MID did when Joel Spingarn started the spy network in the Black community. All Black Americans were viewed as potential enemies of their own country. In 1918 and during World War II, Blacks were suspected of being potential agents of Germany. In the 1960s, Blacks, especially the leaders of the SCLC, were regarded as Communist subversives.

In 1963, President Kennedy and his brother Bobby, the U.S. Attorney General, believed that King's anti–Vietnam War posture and his civil rights marches alienated Whites and threatened Kennedy's reelection chances. In *Racial Matters*, O'Reilly writes that "the Kennedys were active on another civil rights front, spying on the movement through the FBI."[16] It was no secret that Kennedy's successor, Lyndon Johnson, hated King for his suspected Communist ties and opposition to the Vietnam War, and that LBJ "craved" titillating gossip on Black people[17] and "black-scare stories"[18] supplied to him by the FBI, according to O'Reilly.

In October 1967, the FBI started a massive Ghetto Informant Program. Within a year, 3,248 Blacks would sign up.[19] That same year, President Johnson's press secretary George Christian said that he had contacted Black columnist Carl Rowan, a former member of the Johnson administration, who, Christian concluded, "wants to take out after King,"[20] according to author David Garrow in *Bearing the Cross*.

Garrow wrote, "Several days later, Rowan made good on his vow with a column. . . ."[21] In the column, Rowan accused

King of being consumed with self-importance. He implied that King was under Communist influence. Earlier, in 1964, FBI Director J. Edgar Hoover had circulated a highly explosive document called "Communism and the Negro Movement."[22] This top-secret report was soon in the hands of the Army, home of the military intelligence operation that formally organized government spying on Black America in 1917–1918.[23]

Rowan repeated the King-Communist link in the September 1967 *Reader's Digest,* one of the largest circulated journals in the world. "Sinister Murmurings" was the subtitle of Rowan's article, in which he wrote that King was associating with, or influenced by, "enemies of the United States."[24]

In one particularly ominous passage in the *Digest* article, Rowan wrote that King "has become persona non grata to Lyndon Johnson, a fact that he may consider of no consequence," and then the columnist warned King that "the murmurings are likely to produce powerfully hostile reactions."[25]

In the 1980s, Rowan attempted to rewrite history by portraying himself as having been an ally of the slain civil rights champion. In his book *Breaking Barriers,* published in 1991, Rowan writes that he believes "that some in the U.S. government had put out a contract to 'neutralize' the black preacher. . . ."[26] Rowan, who was the first Black to sit on the National Security Council, said his conclusion is "based on all the intelligence data that I had seen while at USIA."[27]

In his book, Rowan lays the blame on his former Communist-fighting ally J. Edgar Hoover. "I shall go to my grave believing that Hoover, Sullivan, and others in the FBI had a role in silencing the black man they professed to fear, but surely hated."[28] Suggesting that J. Edgar Hoover was

"Mary," the transvestite, and blaming him for masterminding King's death is now fashionable among Rowan's liberal friends, but in spite of his modern-day posturing and breast-beating about raising money for "minority" scholarships, Rowan was no friend of Martin Luther King. According to David Garrow, King himself wrote that Rowan's attacks on him backfired and tarnished Rowan's reputation. Garrow called the attack a "red-baiting smear."[29]

In 1967, Rowan's acrimonious attacks on King were enough to provoke the anger of Andrew Young, a King aide who would become the U.S. Ambassador to the United Nations. Young was quoted in the August 28, 1967, *New York Times* as calling Rowan "the worst kind of sophisticated Uncle Tom."

Stoking the Flames

Stokely Carmichael, the flamboyant Black radical Marxist, and his inflammatory and taunting behavior may have pushed military intelligence analysts to conclude that King was ready to light a match to every city in America. In meetings with King, he incessantly called for violent action against the government. Carmichael had to know he was under military surveillance, which he was.

The truth may never be known, but it is a pretty good guess that the militant Carmichael's access to King may have provoked the military and given its leaders motive to order the assassination of King as a potential threat to the security of the country.

Only the President of the United States had the power to order such an assassination. A recently released letter by

Jack Ruby, the man who shot Lee Harvey Oswald, has added fuel to that possibility. In the letter, Ruby allegedly said that then Vice President Lyndon Johnson was a part of the plot to kill President John F. Kennedy and passed information on to Oswald about Kennedy's trip to Dallas.[30] Other sources have also linked LBJ to the assassination. Rowan's warning of Johnson's wrath toward King may have been prophetic. Or somehow a rogue under his command group of the Pentagon's secret agents (now called the Defense Hummit Service) or some CIA rogue agentry could have acted without the President's approval.

According to Thompkins's research, the King march that turned violent on March 28 in Memphis convinced someone in the government that the civil rights champion had it within his power to incite violent insurrection. Undercover agents taped Rap Brown, Carmichael, and King at the Pitts Motor Hotel in Washington and heard Carmichael say, "They bring the army, we fight the fuckers. We got guns,"[31] according to documents the *Commercial Appeal* reporter uncovered.

Further panic set in when Carmichael's close association with King convinced military intelligence that he had bought into Carmichael's urgings to declare war on the government of the United States—a war the nervous military did not feel it could win, considering its dearth of domestic combat-ready troops. Carmichael had said that he would lead an armed insurrection in the United States: "We are organizing urban guerrillas in the United States . . . to bring the collapse of capitalism and imperialism. . . . We are not waiting for them to kill us. We kill first. . . . We must make vengeance against the leaders of the United States. . . . We have no alternative but to use aggressive armed violence."[32]

He also had threatened to kill President Lyndon Johnson, Secretary of Defense Robert McNamara, and Secretary of State Dean Rusk, according to the *Commercial Appeal*.

While King was promoting his upcoming Poor People's campaign, Carmichael was photographed in Havana with a radical Communist revolutionary in late July. In January 1968, he was in Hanoi. Carmichael was linked also to avowed Communist revolutionaries in Cuba and China, and intelligence analysts believed these connections would supply, or were supplying, him with money and guns.

In February, the man of violence met with the man of peace. Undercover agents heard Carmichael tell King: "We got guns."[33] Two months later, on April 4, only days before his planned Poor People's campaign, Martin Luther King was assassinated on the balcony of the Lorraine Motel in Memphis.

On the day King was assassinated, "Eight Green Beret soldiers from an 'Operation Detachment Alpha Team' were in Memphis carrying out an unknown mission,"[34] according to Thompkins, who noted that such units usually contained twelve members. There was also a group of Green Berets from the 20th Special Forces Group in town, he reported. The 20th was notorious for being the dumping ground for "crazy guys" from Vietnam Special Forces "who had worked in murky clandestine operations with the CIA, the Special Operations Group (SOG) or the top secret Detachment B-57,"[35] according to Thompkins.

We know that the bullet that killed King was delivered by a marksman, an expert human guidance system who knew how to mangle his victim's brains from a distance. The weapon used to kill King was a .30-06 rifle, the type the Army had supplied to local police forces, including the

Memphis police department. The rifle allegedly used to kill King was found neatly packaged in a box in a doorway near the apartment house of James Earl Ray, who was convicted of the killing. Oddly, the FBI never conducted a ballistic test on the bullet that allegedly killed King or the rifle that the police say killed King. As a result, no specific rifle was ever identified as the murder weapon by the FBI or the House select committee investigating the murder.

King had returned to Memphis the day before and was monitored by Army agents from the 111th Military Intelligence Group from a sedan filled with electronic equipment, Thompkins reported. The agents also were receiving information on King's movements from informants in his party and/or on the planning committee, according to the *Commercial Appeal*.[36]

Conspiracy of Violence

When the riots started in Washington, D.C., following King's death, an armed Carmichael was among the leaders. He told the *Washington Post* that he wanted to organize this violence into a force to destroy America. The violence pleased Carmichael and he saw it as a symbol of the end of Martin Luther King's nonviolent movement and the beginning of an inevitable conflagration—"the higher the flames reach, the more I like it."[37]

After years of surveillance of Carmichael and his explicit threats of anarchy and violence, the U.S. government never charged him with a crime and never harmed him, at least to my knowledge. He has been allowed to live peacefully in the African country of his choice and to travel freely in and out of the United States. Given its history of dirty tricks against foreign leaders and domestic radicals, this seems to be a deviation from our government's usual practices. After all, this same U.S. government spent millions of dollars and hours of extremely expensive surveillance to arrest the mayor of Washington, D.C., Marion Barry, for smoking crack cocaine in a hotel room while trying to solicit sex.

Carmichael is now a member of the All-African People's Revolutionary Party and is called Kwame Toure. He reportedly organizes socialist chapters among Black college students who have no personal memories of the King era. Carmichael, apparently, is still waiting for a violent revolution that most Black Americans don't believe in. Riots are usually linked by liberal thinkers to unemployment. A conservative once reminded me that "people on the right riot also. If welfare and liberalism get out of hand, look for one." But the kind of low-intensity warfare that Carmichael fantasizes about is most likely to come from a growing polarization of haves and have-nots.

Renaissance Fraud

From the moment that the Rev. Martin Luther King was shot on the Memphis hotel balcony, conspiracy theories arose. But in the history of Black and White relationships in this country, there are many other incidents in which insidious forces have endeavored to thwart self-sufficiency and economic development in the Black community.

In most history books, the Harlem Renaissance of the 1920s is heralded as a triumphant period in which Black culture was finally and fully recognized. Let me offer you a different perspective: It was an artistic and political fraud. The Harlem Renaissance was a smoke screen in which Black art and artists were used by elitist Blacks and their White manipulators to divert the Black masses from their growing efforts to become self-reliant.

This theory that the Harlem Renaissance was a counterfeit cultural movement was developed by Black historian David Levering Lewis. He described it as a "forced phenomenon organized by the Black and White leaders of the Black protest movement. . . ."[38] Lewis has noted that at the time many Black advocates of self-sufficiency such as Marcus Garvey saw through a "renaissance" that was largely staged and manipulated by outside forces interested in distracting Blacks from far more important matters. Garvey wrote that White liberals in the NAACP were "disarming, dis-serving, dis-ambitioning and fooling the Negro to death."[39]

Garvey and others recognized that the Harlem Renaissance was mostly subsidized by Jewish liberals, whether misguided or purposeful. "Nothing could have seemed to most educated Afro-Americans more impractical as a means of improving racial standing in the 1920s than writing poetry and novels or painting,"[40] Lewis has observed.

Even the elitest assimilationist W.E.B. Du Bois, who hated Garvey and worked to discredit him, came to realize just how corrupt and vulgar Black elitism and its Renaissance charade had become. Lewis recounts Du Bois's exasperated recantation: "Although his own magazine had helped promote the [Harlem Renaissance] movement, Du Bois came to disapprove of a racial program offering poetry in the place of politics and Broadway musicals in the place of jobs."[41]

Many of its participants later considered the Harlem Renaissance more of a rich White man's trendy adventure into

Black bohemia. Harlem writer Claude McKay lamented that "the Harlem Renaissance movement of the artistic '20's was really inspired and kept alive by the interest and presence of white bohemians. It faded out when they became tired of the new plaything."[42]

The Harlem Renaissance was designed as a fatal distraction for Blacks to be diverted from the fact that the very sponsors of the Harlem Renaissance had just destroyed the head of the largest Black mass movement in history, Marcus Garvey, and the most viable vehicle that Blacks had, no matter how immature, for Black economic development.

However, the aristocrats' fatuity cut them off from their own reality. The Harlem Renaissance demonstrated the extremes the Talented Tenth (ten thousand of ten million Blacks in 1920) went to in using rank-and-file Blacks as a footstool for the advancement of *certain* colored people. They felt entitled to this privilege because they descended from either free Blacks or Whites or both. They developed separate institutions such as the American Negro Academy, headed at one time by Du Bois, for intellectuals and exclusive social clubs like the Mu-So-Lit in Washington, D.C., the Agora in Nashville, and the Crescent in Cincinnati.

Out of this elitism grew the conceptual bias for noneconomic socialism: the belief that the vast majority of the Black population was not biologically (because they were dark-skinned) or environmentally (because they lived in squalor) capable of competing economically and educationally with White people.

Du Bois and his Talented Tenth minions were guided by this philosophy as they assumed command of the Black community. Naturally, as noneconomic socialism dictates, they advocated dependence on White people and the govern-

ment. This antieconomic policy of "the better class Ne-
groes" worked successfully for the well-educated Talented
Tenth, which had four or five generations of freedom and
often a college education at the turn of the twentieth cen-
tury. Failure, of course, became an inevitability for the com-
mon people; it devastated and underdeveloped the largely
down-and-out Black masses who had been freed from slavery
for only one generation or less. In effect, this treachery of
elitism and noneconomic socialism reenslaved Black people.
And it does to this very day.

The Toll of Conspiracy

The lessons to be drawn from these sorry episodes in Black
history are many. What Black mainstream leaders did not
learn in the early part of this century they have not yet
learned. They are still elitist assimilationists in concept, and
socialistic in function. The harm of integration and the fail-
ure to seriously embrace economic self-sufficiency are les-
sons that the BUM has still not grasped and certainly has
not truly embraced.

The Black masses have suffered for almost one hundred
years from demagoguery. History has exposed the Harlem
Renaissance as a diversionary hoax perpetuated by Black
leaders (the BUM) against members of their own community.
The primary motive was to hide their complicity in the de-
struction of a Black vision of economic self-sufficiency.

The most brilliant summation of this Black tragedy is
that of historian Lewis, who understands the origins of a
failed Black leadership. He offers that the Harlem Renais-

sance "literally took place in rented space—in a Harlem they did not own."[43]

Lewis and other perceptive historians have pointed out that Black aristocrats "missed the significance of the butcher and tailor shops, the sweatshop, the pawnshops, and the liquor stores. . . . It is not surprising, then, that on those few occasions in the early 1930s when the Talented Tenth mobilized the masses to protest economic discrimination, its specific demands were usually for middle-class advancement."[44]

Lewis concluded: "The Harlem Renaissance bubble would soon go flat, but the assimilationist values and goals of the Talented Tenth would be perpetuated in civil rights strategies in which the emphasis remained on court cases, contracts, contacts, and culture."[45]

The conspiracies against Black America have taken a heavy toll. We have lost our leaders and we have lost our vision. Unless Blacks revive themselves, the inner cities will consist of drug-addled criminals surrounding a coterie of abandoned, single, pregnant women—all uneducated, untrained, and useless in a world of high technology. All of the cotton has been picked. And for the first time in the history of the United States, Blacks are not needed to sustain the economy. Unless a new Black leadership emerges and halts the vicious cycle set in motion at the turn of the century by noneconomic socialism, integrationism, assimilationism, and self-hatred, the entire nation is in peril.

FEAR OF GENOCIDE

According to Murray, blacks aren't inferior, just dumber.
—review of *The Bell Curve* by RICHARD J. HERRNSTEIN and
CHARLES MURRAY

When I began writing this book, I was tentative about bringing up the subject of genocide and its more subtle form, *triage*, out of fear of being labeled paranoid or alarmist by those not familiar with the history of conspiracies against the Black community. No more. Not after the mass media made my case for me with its rush to hail the publication of *The Bell Curve*, which the *New York Times* editorialists called "a flame-throwing treatise on race, class and intelligence." The *Times* noted that the book has "a grisly thesis: IQ, largely inherited and intractable, dictates an individual's success—an economic death knell for much of America's black population."[1]

While Americans in general blame their pessimism on the fear of economic and social decline, a growing number of Blacks share the belief that there is a plot to promote racial warfare in order to keep their race marginalized, or, most frightening of all, to exterminate Blacks altogether. In this chapter, I will examine the inflammatory but persistent issue of racial genocide in this country. And I will explore the deep-seated, if not openly expressed, fear among Blacks that the ultimate goal of White Americans is to wipe out the entire African-American community. As one man said on my television program, "The logical conclusion of racism is genocide." The thought of this may be so repulsive to some people that they deny that it is even a possibility, but to do that is to deny history. Bosnia and Rwanda are the latest reminders that the world has not abandoned "ethnic cleansing" as a geopolitical solution or weapon.

Let me note that I think a campaign of genocide against Black Americans is a highly unlikely scenario, for political as well as other reasons, unless there is some radical shift in this country to an authoritarian government. On the other hand, I believe that a more insidious threat exists to Blacks in America. As a historically marginalized people, this sociological minority is more vulnerable than any other group to a more subtle campaign of extinction, a form of selective elimination known as triage.

The term *triage* first came into common use in this country in reference to a system of emergency medical treatment developed by the French and implemented by American medical units during the Vietnam War. It is defined as the sorting of and allocation of treatment to patients and especially battle and disaster victims according to a system of priorities designed to maximize the number of survivors.

On the battlefield, the priority is to save those who can be saved, and not to waste precious resources on either those who are not in danger of dying or those who are already too far gone. In a societal sense, triage takes on more malevolent tones. You might think of it as the "rational" approach to genocide, culling those who are the least productive, or beyond redemption.

A precursor to *The Bell Curve*, and the book that I believe is the seminal work on social triage, was *Report from Iron Mountain*, written by Leonard Lewin. It was a publishing sensation in 1967, just as *The Bell Curve* became twenty-seven years later. But when *Iron Mountain* was published, no one knew if it was fiction or nonfiction, and the author wasn't telling then. Both *Iron Mountain* and Lewin's following book, *Triage*, were fiction, but both seem even more chilling today, considering the increasingly volatile racial climate and serious consideration of "high-tech" reservations[2] for Blacks by *The Bell Curve*'s authors.

In the ominous *Iron Mountain*, Lewin wrote of a team of scholars and scientists working for the U.S. government, and he had them recommend that slavery be reinstituted in this country.[3] In *Triage*, which was more clearly presented as fiction, Lewin tells how the United States might eliminate marginal and unproductive population groups by denying them services and slowly cutting them off.[4] Today, Lewin's *Triage* scenario sounds like Charles Murray's view of Head Start.

Not surprisingly, both of Lewin's books have become increasingly popular among White supremacists in recent years—a fact that outrages the author, or so he told me. Both *Iron Mountain* and *Triage* offer blueprints for a sophisticated and diabolical culling of undesirable population groups.

If Lewin's books, the more recent *Bell Curve*, and similar books and articles don't incite paranoia and alarm in the Black community, then Blacks could rightfully be accused of being as mentally stunted as *The Bell Curve*'s authors imply. These authors do demonstrate that their "White" IQs do not confer character.

In *The Bell Curve*, Murray and coauthor Richard Herrnstein, a Harvard psychologist who died just before the book was published, infer that welfare programs, affirmative action, Head Start, and any other efforts to help elevate the lives of the American underclass have been a waste of time because Blacks are genetically flawed. In fact, the unstated implication is that Blacks in general are a waste of time. And that, of course, officially marginalizes an entire race of people. This makes Blacks candidates for eventual extermination, at least to anyone who has studied the history of genocide or triage.

There is nothing new in Herrnstein and Murray's book. And there is nothing unusual about the news it delivers. There is also nothing valid about it. Studies of the Black African immigrant population in Great Britain have shown that they have higher educational achievement than any other population segment in that country—a fact that Herrnstein and Murray seemed to have overlooked.

FINDS TOUCH OF AFRICA IN 28 MILLION WHITES read the headline in the *Sunday News* of June 15, 1958. Historian J. A. Rogers dug up an Associated Press story in the *Sunday News* that reported on an article in the *Ohio Journal of Science*. It claimed that 21 percent, or approximately forty million White Americans today have African ancestors.[5] Without racial purity, how do we measure racial intelligence?

What really troubles me more than anything else about

The Bell Curve is its timing. White people are under stress, and Black people don't need to have them aggravated. Because of the runaway national debt, Whites' standard of living is threatened, and they only have to look across the office or down the street to find someone to blame. We all know what happens when the people with the strength and the numbers come under stress. The laws of survival kick in. The strong go looking to bash the weak. And the weak had better go looking for a good place to hide.

The Bell Curve's authors recognize this, even as they encourage it. In the book, they predict that "racism will re-emerge in a new and more virulent form." The authors offer that "instead of the candor and realism about race that is so urgently needed, the nation will be faced with racial divisiveness and hostility that is as great as, or greater than, America experienced before the civil rights movement. . . . If it were to happen, all the scenarios for the custodial state would be more unpleasant—more vicious—than anyone can now imagine."[6] It isn't difficult to imagine what the authors are suggesting.

High-Tech Reservation

By "custodial state," *The Bell Curve*'s authors say they mean "a high-tech and more lavish version of the Indian reservation for some substantial minority of the nation's population, while the rest of America tries to go about its business." In it harshest forms, they write, "the solutions will become more and more totalitarian."[7]

As the authors imply, Whites are fed up. And Blacks are in trouble. Today they are talking about a high-tech "reservation." Back a few hundred years, it was simply called "slavery." The White would-be masters are already making their case. Although presently a fringe group, the White intellectual supremacists are growing in direct proportion with the country's economic decline. The latest books reflect the widespread belief that America's underclass is growing ever bigger, and ever more threatened. "It is difficult to imagine the United States preserving its heritage of individualism, equal rights before the law, free people running their own lives, once it is accepted that a significant part of the population must be made permanent wards of the state."[8]

Later in his review of *The Bell Curve,* *New York Times* science reporter Malcolm Brown notes that "unless future accommodations between ethnic groups lead to a more harmonious social structure, Herrnstein and Murray say, the potential for racial hatred seems enormous."[9] This sort of thinking holds that social programs have failed and the Black underclass has only grown larger because Blacks are intellectually inferior because of genetic makeup. There is nothing that can be done for this genetically inferior class of people, and so nothing should be done for them—as this revived line of eugenic thought goes. "Their implication is that blacks are trapped at the bottom of society,"[10] the *Times* editorial said.

As the *Times*'s own E. J. Dionne, Jr., noted in a rebuttal editorial, the arguments presented in *The Bell Curve* are rather old-hat, and therefore should not have been big news, except, given the current level of frustration over failed social policies, any explanation is welcomed. This book, Dionne noted, is nothing more than "a flashy repackaging of a repeatedly discredited fashion in American life. Whenever we are exhausted with reform, we shrug our shoulders and say, 'There's nothing we can do for that poor guy in the street.' Thus was pseudo-science about racial differences used to justify the end of Reconstruction and the reimposition of a segregated caste system on the American South.''[11]

Critics have noted that the Herrnstein-Murray book and others like it are not written for the sake of any science, they are created for political reasons. I am not disturbed so much by the book as I am by the fact that in the current climate so many others accept it as news, or as science.

There is every reason to believe that Black people are still easy targets for genocide. Black people know that, and they often talk about it among themselves. Generally, however, we are duplicitous when we discuss the subject around Whites. With good reason. We don't want to feed the flames.

What White Folks Don't Hear from Blacks

A Black physician whom I know well was quoted in the *New York Times* as telling a reporter that when Blacks spread rumors about genocide, they victimize their own people by making them feel powerless and thereby provide poor Blacks with another excuse for not helping themselves. That same Black man told me later in a private discussion with several other Blacks present that the CIA was flooding the streets of the inner cities with drugs to kill Blacks.

When out of White earshot, this prominent doctor freely offered his belief that racism and government conspiracy are at the core of the social crisis among Blacks. He is not a hypocrite, though he may seem one. He is merely exhibiting the duality reflected in the lives of most Blacks living in a world dominated by Whites.

Fears of genocide run deep among Blacks even today. Another Black intellectual drove this lesson home to me once after I informed him that his own particular conspiracy theory had no basis in fact. "Just because you don't think ain't nobody standin' behind you don't mean they ain't," said this very well-educated, articulate scholar. He intentionally affected a Black rural dialect to emphasize that this suspicion is deeply rooted in our culture.

He was reminding me that any Black, especially a Black man, who is not highly suspicious is crazy, especially given our experience in this country. While it is imperative to remain in touch with reality, it's important that we know and understand history.

As animals learn adaptive patterns, so do humans. Black paranoia, to some acceptable extent, is a learned response to continual abuse. In my opinion, this makes the clinical definition of paranoia as a symptom of mental breakdown— a loss of touch with reality—subject to some cultural allowances. Looking at it from another perspective in an appropriate context, Black paranoia, if I may call it that, could be an expression of an advanced intuition. To be sure, whatever else Blacks are not, they are survivors, and I don't think that is accidental.

Rumor can be a weapon of sorts in a group under domination. In *I Heard It Through the Grapevine*, author Patricia Turner offers that Blacks spread negative rumors about products and companies that they feel exploit them as a form of economic retaliation. In political matters such as "AIDS" and the government, Blacks put rumors of impending disaster on the drum as a form of warning and mobilization. The messages go exclusively from one Black to another. Professor Turner explains that it is not surprising that this area of Black communications is unknown to most non-Black Americans, because folklorists, most of whom are White, ignore it. Turner's survey found that her Black subjects generally identified the perpetrator of these conspiracies as only "the government or the powers that be,"[12] but that 10 percent fingered the CIA specifically.

Those in that 10 percent of the survey may have read *The Secret Team: The CIA and Its Allies in Control of the United States and the World,* which was written by retired U.S. Army Colonel L. Fletcher Prouty. He served as the focal point officer between the CIA and the Department of Defense from 1955 through 1963, and in that position was privy to information few outside the intelligence community ever see. Prouty reports that a "High Cabal" of rogue CIA operatives has spread its tentacles into every important function of the government and business world and that the cabal's members operate outside the control of the President, Congress, and the military.[13]

The major media were named by some of Turner's respondents as co-conspirators because they "are advancing the government's conspiracy by bombing the public with images of drug-pushing and drug-abusing blacks,"[14] according to Turner.

In American society, there is a cultural predisposition to believe that Black men and violence are synonymous, and this translates to serious trouble for Blacks when Whites come under stress.

Although the media and opinion polls report on a general White fear of Blacks, ironically, Blacks are genuinely afraid of Whites, especially Whites under stress—the same sort of stress likely to increase during economic hard times. Blacks should also be strongly concerned about the faltering economy and moral decline for the same reasons. Among the historically rooted themes that cropped up during Turner's research on Black rumor-mongering were fears of "conspiracy, contamination, cannibalism, and castration—perceived threats to individual black bodies, which are then translated into animosity toward the race as a whole,"[15] Turner writes.

As I documented in the previous chapter, there is ample historical background for suspicion, and a certain healthy paranoia, among Blacks even today. No one blames the Jews for remaining vigilant about new and potential threats to their well-being. This is not to say that every Black, whether a gang member or a public official, who is investigated and charged with a crime is the victim of a conspiracy to undermine all Blacks. Indeed, the ones quickest to play the race card tend to be the guiltiest.

Responsibility in Blacks and Whites

Many Blacks fear that genocide is a real possibility because they recognize, on some level, just how hopeless their condition has become. And they recognize, perhaps intuitively, that it will only get worse. They see the poor quality and lack of focus in their elitist Black leadership. They see the counterproductive behavior of a growing "gangsta" subculture among Black youths. They see the emergence of media-conscious Black hate-mongers. They see White denial of the fact that our problems—Black and White—are intertwined with the nation's own destiny. Most of all, they see a knowledge sector and a new world of high technology being built without them in cyberspace.

Black anxieties are heightened when they hear demands for an end to the very problems that have become euphemisms for Blacks—crime, illegitimacy, poor values, poor schools, and, most of all, violence. At best, Blacks feel ambivalent about most proposed solutions to societal problems because they realize that their racial community is recognized as America's socioeconomic Achilles' heel. Eventually, Blacks fear, conditions in this country will reach a breaking point, and the ax will come down—on them.

I believe that if Blacks found and brought about solutions to their own problems, it would not only give a boost to the lives of Blacks, it might also jump-start the entire nation and move our United States back into its leadership role in the world. Blacks and Whites alike understand that if America is to remain a viable nation, we must solve these problems.

If our national economy goes bankrupt, we all know which racial group would be judged too far gone to be saved. Already, Black Americans have been systematically cut off from economic activity and economically and socially marginalized by racism. When high rates of unemployment, a lack of productivity, and an assortment of societal ills— crime, drugs, disease—are blamed on Black "racial traits," it naturally follows that the majority group would increasingly see Blacks as a drain overall on society.

Just recently, this sort of highly rational thinking has been applied in China, where government officials instituted a ban on marriages that might produce children with defects. The totalitarian Chinese government's stated goal is to upgrade the quality of the population. This is known as eugenics, a process that often involves genocide or triage.

The Failure of Preference Programs

Recent structural changes in the American economy and a downturn in racial relations have hastened the process of marginalization, pushing Black Americans farther to the outer reaches of society, where it becomes easier to nudge them over the edge. Democratic Party propaganda holds that relations between Blacks and Whites have soured because Republicans were in the White House between 1980 and 1992. Closer to the truth is the explanation that, in addition to being hurt economically by recession, Blacks have lost economic and social standing because of a failed strategy among their own leaders and their White liberals "friends," notwithstanding an overall shift of wealth away from the poor and middle class to the wealthy.

The constant quest for preferences by Black leaders has resulted in the perception among the majority of Whites that Blacks want to move up by shoving Whites back—a dangerous perception, even if it is not true.

Politically, then, the strategies of the elitist Black Unaccountable Machine (BUM) and White liberals have backfired. As a result, every new Black job created by preference programs was perceived as a net loss to a potential White worker. This has provoked racial confrontation.

Even well-trained and well-educated Blacks have become stigmatized as incompetents who achieved only because they got help. Instead of helping desegregate society, these preference programs have served largely to further marginalize Blacks and alienate them from the general population. It is impossible to explain logically to a White worker that his or her unemployment makes this a better country. The reality is that Blacks are desperate to catch up to Whites economically, and while Whites in general may be all in favor of that, no single White wants it to come at his or her expense. This zero-sum-based policy defines one group as an undesirable scavenger population. In short, Blacks are in a lose-lose situation politically.

Like the Jews in Germany prior to the Holocaust, Blacks in America have lost the public relations battle. By 1994, much of White America had grown so hostile toward Blacks that even when middle-class Blacks lowered the socioeconomic barrier that separated them from middle-class Whites by education and income, opinion polls found that White households still did not want Blacks as neighbors. The fear is that once Blacks get a toehold in a suburban area, more Blacks follow, inspiring more White flight, creating a new Black neighborhood. So much for one-way integration as the panacea for the Black masses.

It is my guess that the White population is not behaving this way because it is racist (although some Whites undoubtedly are) or because it is Republican. It is obvious that most Whites are convinced that any association with Blacks will reduce the quality of their lives. That is an economic assessment that quantifies the Black population as an economic liability. When Whites vote, they want someone—Republican or Democrat—to craft policies to contain the Black population. In modern societies, that is a precarious position for a marginalized group.

For those who have not studied genocide, it may be difficult to comprehend, but genocide tends to be a highly "rational" action, although a very immoral, unethical, and sometimes illegal one. For the past five hundred years, civilization has been guided by a rational approach in which the strong kill the weak in order to remain strong. Throughout human history, lip service has been given to the moral aspect—to the value of human life—but in reality, the struggle for dominance of one person over another, one group over another, or one nation over another is without end.

A Half Million Dead Africans a Year

In her chilling article "The Etiology of Genocides," Barbara Harff lists the major extermination campaigns of the twentieth century. Her list includes up to 1.8 million Armenian victims of the Turks or Kurds in 1915; up to three million Ibos murdered by other Nigerians from 1967 to 1970; up to three million Bengalis massacred by the East Pakistan Army in 1971; up to three million Kampucheans exterminated by the Khmer Rouge between 1975 and 1979; and up to six million Jews and 48,000 Gypsy victims of the Germans between 1941 and 1945.[16]

Harff does not mention the efforts of King Leopold II of Belgium, who fifty years before the Jewish Holocaust reduced the African Congolese population from twenty million to ten million within just two decades. In writing of Leopold's infamous slaughter of Black humanity, Mark Twain noted, "There are many humorous things in the world; among them the White man's notion that he is less savage than other savages."

Genocide grows in a recognizable atmosphere with identifiable warning signs, amid a host of genocide-related events or circumstances. Historically, when mass death is linked to a surplus population, "the chronically unproductive and generally unemployable"[17] are eliminated during times of crisis, according to Professor John K. Roth, who wrote an article in the book *Genocide and the Modern Age*. The genocides enacted by Joseph Stalin and Adolf Hitler were a means to get rid of a surplus population that those dictators saw as a problem. Hitler saw to it that Jews were cut off from economic activity until they were marginalized as chronically unproductive and generally unemployable. Then these "racial traits" were used to make scapegoats of them and to target them as the cause of Germany's economic upheaval before their extermination began.

Dirty Tricks

The history of politically and economically marginalized people in this century shows a record of mass murder that includes genocide in its most extreme form, holocaust, or its less overt form, triage. Totalitarian regimes have been facilitators in that process. Stalin's liquidation of the Russian peasants following the Russian Revolution in 1917 is a historic example of encroaching totalitarianism. A modern American example is the administration of President Richard M. Nixon, who intensified his attacks on his perceived enemies even after he won control of the White House in 1972.

Nixon's administration used "dirty" tactics normally reserved for foreign enemies. He used the CIA against American citizens and practiced bureaucratic terror, especially tax harassment, to keep his opponents on the defensive. Such actions are characteristic of modern totalitarian rulers.

Imagine what kind of America this would be today had some CIA-led burglars not bungled a break-in at the Watergate Apartments. I believe we came very close to a totalitarian rule. And there were more recent, if less grave, echoes of that sort of misuse of power in the Clinton administration's attempts to use the IRS against its enemies in 1994. The conduct of the Clinton administration's HUD and the FDA are equally fascist. To me, it is obvious that whether Republican or Democrat, modern government, even in the world's foremost democracy, still manifests signs of totalitarian tendencies when dealing with perceived enemies if it looks as though the government can get away with it.

In a totalitarian government, no one, Black or White, Democrat or Republican, rich or poor, is safe. In light of our polarized racial history, it is probable that a dramatic and severe rise in violent crime will lead to a more totalitarian leadership at all levels of government. The election of authoritarian types could be the majority's call for a triage criminal system. The "violence initiative" of the National Institutes of Health—the same organization that performed the Tuskegee experiment on Black syphilis victims—is the perfect guise for triage. One leading NIH official left no doubt about his racist orientation when he characterized Black criminals as "monkeys" and recommended early intervention with questionable drugs for young Black men to control their alleged natural criminal tendencies.

The world community, the external brake on national genocide, would perhaps look upon American triage as a long-overdue resolution to a festering domestic problem. A crime-weary and frightened U.S. population would look the other way. The medical community could not be counted on to step in and question the morality of triage or genocide. Hilter's killing camps, it should be noted, were nearly always headed by a medical doctor, often a psychiatrist.

The idea of a coercive scheme to incite racial violence and the maintenance of a hate climate by government agent provocateurs is common to most conspiracy scenarios envisioned by those Blacks who believe genocide is a possibility—and that includes most Blacks, whether they admit to it or not.

More recently, I have been aware of a contemporary movement to entrap Blacks and Whites into a dance of violence. In 1986, I wrote about Whites who viewed a "race war" as inevitable and therefore were preparing themselves psychologically to destroy the young Black men who would precipitate it. I wrote back then:

> A leading financial expert is telling his clients that an impending crisis will precipitate urban civil disobedience and that inner-city attack groups of young Black men will invade nearby suburban areas. To avoid this race war Whites should now move to all-White small towns, he advises.
>
> That financial consultant predicts that this would be a tragedy for America's Black people. The sheer numbers and stockpiles of guns are against them.

There is no doubt that the poverty and frustration in America's Black ghettos are social dynamite. And they could be ignited by an economic collapse or precipitated by any number of unrelated incidents. The role of rumor in race riots cannot be underestimated. Whether inner-city "attack groups" ever existed or did initiate attacks on suburban areas would never be known, just as unfounded reports of dead babies in Detroit in the 1940s fueled that city's worst race riots in history. It is a fact that human carnage has followed the mere anticipation of a race war—and shooting wars are preceded by cultural wars.

Choosing Sides

In a study by Lou Harris and Associates for the National Commission on the Causes and Prevention of Violence, it was determined that if a fringe group of White extremists, or some rogue unit of "rationalists" within government, succeeded in drawing Blacks and Whites into a race war, the outcome would be particularly catastrophic for Blacks. One question included in the survey offered this scenario: "Imagine that the government has just arrested and imprisoned many of the Negroes in your community even though there had been no trouble."[18]

The responses to that scenario, published in a November 1970 *Psychology Today* article, were revealing. The overwhelming majority of White Americans would apparently be good Germans if the government turned to massive racial repression; 18 percent would protest nonviolently and only 9 percent would turn to violence—for a total of 27 percent who would resist. Blacks understandably would be more willing to act; 43 percent would use civil disobedience and one-fourth would attempt counterviolence. This may reflect a pragmatic judgment that if such things came to pass, Blacks would be wiped out if they rebelled.[19]

It's my guess that, with a worsening racial climate since that poll was conducted in 1968, fewer than 27 percent of Whites would offer assistance to Blacks today. Even assuming that the numbers still hold, it means that 73 percent of

Whites would offer no assistance to a "massive racial repression" of Blacks. Moreover, the survey question contained a big "if." It assumed "there had been no trouble." I interpret that to mean, for example, that no Black zealot had sounded the call to kill Whites, and that Blacks had not initiated violence against authorities. If Black-on-White violence had occurred, White support would undoubtedly be less than the 27 percent. To make matters worse, for Blacks at least, Blacks themselves would offer minimal resistance. As mentioned, less than half, 43 percent, would offer nonviolent resistance, and only 24 percent would resist violently. Another 33 percent would offer no resistance of any type.[20]

It is anyone's guess just how far Whites would allow government repression of Blacks to go, or how much violence against Blacks would be tolerated. Only 14 percent of Whites "would fight to defend Blacks from unlawful and unjustified mass imprisonment,"[21] according to the Harris poll. Whatever the outcome, Americans generally regard violence and war as inevitable, the survey revealed.

With triage or genocide offered as solutions to national problems, it would only be a matter of time before they became the "final solutions" for other perceived group problems and value conflicts. After all, America clings to a history of racial animosity, as does the rest of the world.

Something Wrong at the Top

Genocide expert Barbara Harff writes that democratic institutions are not "safeguards against mass excesses."[22] Democratic systems have turned against their own people. She cites the U.S. government's Trail of Tears, the forced marches from Appalachia to Oklahoma that killed thousands of Cherokees. In a modern democracy, the relationship between the leaders and the people is unique—quite different from the German dictatorship responsible for the Holocaust genocide. In a democracy, genocide would probably take the form of "selective neglect" or triage, rather than the more overt slaughter enacted by the Nazis. Even if the leaders intended genocide, public opinion would have to overtly sanction such an action. Therefore, propaganda and media manipulation to shape public opinion are crucial elements, adjuncts, in any modern-day mass excess.

As we near the end of the second millennium, I believe that confronting the issues of genocide and triage—and the greater issue of racism—is the moral equivalent of Whites facing the issue of slavery in the seventeenth century in Thomas Jefferson's America. The triage economics of the twentieth century in which industrialized countries decide which populations in Third World countries live and which die is an awesome moral responsibility. Legal slavery allowed Whites in the New World to become lazy and enjoy a standard of living unheard of in their native Europe. But 56 million indigenous Taino Indians and millions more enslaved Africans died to make possible the warm houses, transportation, fine clothes, and wealth for the European immigrants.

The Avenging Deity

This spiritual anarchy certainly bothered the third President of the United States, the author of the Declaration of Independence. In 1781–82, Thomas Jefferson, a man known for his "overriding concern" for free Whites and his hypocritical indifference to Blacks, unburdened himself fully in his *Notes on Virginia*. In the only book he ever wrote, the slave owner reluctantly published an impassioned indictment of slavery. Yet he outdid himself in his ham-fisted expression of "his doubts about the innate intelligence of blacks,"[23] historian John Chester Miller theorizes in his examination entitled *The Wolf by the Ears*.

The fact that Jefferson did not intend his soul-searchings to be published, at least not under his name, lends greater significance to his moral concerns. In effect, he was telling the White citizens that "they were living under a form of government that violated some of the basic principles of the American Revolution and that their minds and hearts were being brutalized by the presence of slavery."[24] It was literally a journey into the souls of White Americans.

Thomas Jefferson's dilemma in 1771 and White America's unresolved dilemma in the late twentieth century are the same. While it appears that the social and economic crisis of Blacks is the issue, in reality, it's the souls of White people being judged by Jefferson's "avenging Deity." That's what drove Jefferson to write the lengthy indictment of King George III in his draft of the Declaration of Independence and what caused him to "unburden" himself on the morality of slavery in a book he planned to keep secret.[25]

That may have been enough to qualify Jefferson, who, according to Miller, feared "the avenging Deity," as America's moral policeman. My guess is that Jefferson was being consumed by his fear of America's spiritual anarchy and literally going to hell, where, ironically, he believed there were no White people or Black people—just souls who suffered in the eternal fires of damnation. To his credit and undoubtedly out of his religious convictions, Jefferson wrote in his secret book that the "ownership of human beings fostered only cruelty, false pride, tyranny, and mindless brutality—the most uncivilized behavior of which man was capable except in time of war."[26]

He concluded that America's system of slavery, based on White supremacy, would "serve only to degrade the slave and debauch the morals of the master." Miller adds: "He believed that Blacks could endure slavery and still emerge with their moral sense intact, whereas Whites were utterly demoralized by the intoxicating sense of power it engendered."[27]

Whatever Jefferson's shortcomings were as a human being, he did have a moral compass. His salvation was in the realization that the "master" was indeed the slave of his or her own greed and passions, the evil of his or her lower self, and that God had allowed slavers the illusion of power that had only one purpose: to corrupt the soul. "Power corrupts; absolute power corrupts absolutely," as Lord Acton put it.

A Genocidal History

The concept of Black genocide has reared its ugly head throughout our own country's history. James Callender, the man who publicly accused Thomas Jefferson of making teen-aged slave Sally Hemings serve as his mistress and of fathering "mulattoes," wanted a pure race of White people. In order to control the spread of "Congo Harems," Callender predicted "a series of massacres" that would result "in the utter extinction of both blacks and mulattoes."[28] Eliminating what he termed the evil mulattoes and the miscegenation was the only solution to keeping Virginia "habitable" for White people.

The idea of this sort of Final Solution for Black Americans has not been lost on racists today; a few are already organized and armed for that purpose. The idea of mass extermination is repugnant to the Judeo-Christian ethic and the Jeffersonian consciences of the majority. But just as the "good Germans" winked at Hitler's genocidal killing machine, one wonders whether this modern American dilemma has unconsciously conditioned Jefferson's "virtuous, virginal and moral" descendants to also look the other way if official sources resort to rational, drastic measures. Once again, we are approaching a time when, I am afraid, White America must decide whether to solve its racial dilemma productively or extinguish its perpetual race problem at great moral risk of offending the "avenging Diety."

Moral Dilemma of the Majority

The movie *A Few Good Men* dramatically brought into focus the moral dilemma of the majority in a democratic society. In the film, it was impossible to distinguish between the symbols of good and evil, and it was easy to identify emotionally with both protagonist and antagonist, because we are all a little of both. Someone, the theme suggested, has to stand on the wall—do the dirty work—so the rest of us can sleep at night and enjoy the material benefits of our collective immorality. In the movie's dramatic climax, the righteous young military prosecutor (Tom Cruise) demanded to know the cynical colonel's motive for breaking the law, society's social contract.

Symbolically, the young lieutenant represented a mixture of societal innocence and selfishness, the perfected image we project into the mirror and the lie that we need to tell ourselves in order to keep denial intact. The Cruise character also symbolizes the national soul—the soul that Thomas Jefferson worried about White Americans losing.

Jack Nicholson's colonel, on the other hand, was a veteran of dirty tricks in a shadowy world of deceit where cheating and murder are justified as rational choices. He protected society's best material interests. He symbolized the rational self, always at war with a higher self. Their dialogue symbolizes the structural underpinning of White America's moral dilemma: In symbolic form, the higher moral self challenges its lower self.

When Cruise demands answers and the truth from Nicholson, the colonel bristles, "You can't handle the truth." The veteran military man played by Nicholson then lets fly with his contempt for the young lawyer and any others who dare challenge his judgment or methods while living safely and sleeping "under the blankets of the very freedom that I provide.

"I would rather you just said thank you and went on your way. Otherwise I suggest you pick up a weapon and stand a post,"[29] Nicholson growls.

Can America ever bring itself to "pick up a weapon" to solve its racial problem? Is there anyone to stand the post of morality? That is White America's dilemma, because it is the majority group. The credit or blame for a viable resolution or a mass extermination should go directly to America's White population. As a representative democracy, the majority wins. Therefore, American morality is largely White America's morality and White public opinion is generally White America's perception of issues and events.

The Danger of Hate Talk

Everything has gone to hell for poor Blacks, a totally dispirited group that has been abandoned by all. They are helpless and broken, and a belief in a conspiracy to wipe them out could be a death wish, or perhaps even a cry for help.

My overriding concern, and I cannot overstate it, is the incendiary nature of the hateful rhetoric spewing from fringe groups of the Black community, whose venom threatens the entire Black population. Ignorant of history or perhaps inspired by history's darkest aspect—mass extermination—these zealots play with fire at our collective expense. The great majority of thoughtful Blacks find this racial intolerance among their own people repugnant. I predict that to exacerbate this dilemma, as economic conditions worsen for the lower stratum of Blacks, the belief in a White conspiracy of Black genocide will spread further and deeper.

Unfortunately, the most desperate increasingly will demand new fanatical leaders who sound like this: "I want to be one of the flamethrowers of God, break White folks' backs. I want to give you hell all the way to your [Whites'] graves. I ain't scared to die and I'm ready to kill."[30]

Those are the words of Khalid Muhammad, the former chief spokesman for Minister Louis Farrakhan. Muhammad was speaking to an enthusiastic Black audience in Brooklyn. It was not his first verbal dare to White America. He told a Washington, D.C., rally in April 1986 that if any attempt to imprison or harm Farrakhan took place, "the people will burn this country to the ground."

Muhammad's hyperbole is greatly exaggerated and his influence substantially less than popularly imagined. But the number of Blacks who think, and, more important, feel, as Muhammad does is growing. If they are incited to violence, Blacks at large would be massacred, of course. But logic and facts do not get in the way of poor judgment or zealots who are "flamethrowers of God."

"It's as though White America is sleepwalking on the edge of a volcano of ethnic and racial differences,"[31] said Sanford Cloud, Jr., president of the National Conference of Christians and Jews, in a *Washington Post* interview. The biggest threat is that White America will suddenly awaken and respond to these increased tensions spawned by Black "gangsta" entertainers and fanatics who miscalculate the danger that an aroused majority historically poses to a despised racial minority.

Barbara Harff concludes that all of the genocide cases she has studied "were preceded by challenges to the dominant power strata" and accelerated by "polarization."[32] In that historic context, think about the emergence of in-your-face Black threats, such as a rapper's call to kill "one White a week." There is an increasingly large audience attracted to this theater of the absurd. For example, after identifying the devil as the White race, Muhammad asked his Black audience in Brooklyn to support the Long Island Railroad mass murderer Colin Ferguson, who aimed only at Whites with his 9mm weapon. "I thank God for Colin Ferguson," said Muhammad, adding, "Colin Ferguson was commanded by God. God sends earthquakes. God sends hurricanes."[33]

Some might argue that God sends genocide to eliminate

Black people who are considered "redundant" and a threat to the majority.

I believe that America is a powder keg. Armies of White "survivalists" and militias are polishing up their rifles in anticipation of a violent showdown with Blacks. But they are not alone. Many mainstream Whites are increasingly resentful of Blacks. They see underclass Blacks as drug-pushing, car-jacking, gang-banging wild beasts. And they view middle-class Blacks as threats in the workplace and in their neighborhoods.

To the Black "gangstas" and "flamethrowers" and their would-be recruits, I offer this advice: "You are playing with fire." And the militias are poised to throw gasoline upon the flames.

The Self-healing Process

Reconciliation and healing are the prescription for America's intractable race problem. But healing is made difficult by our inability to put events and feelings in a timely perspective. For all intents and purposes, there is no such thing as a past or a future—only the present—which is why the legacy of slavery will not go away unless we confront it with a healing process in the present.

Contemporary White Americans are not morally or otherwise responsible for slavery or the failings of their ancestors, any more than the descendants of Africans are responsible for the Africans who supported the European slave trade of their own people. White people are morally responsible for allowing the descendants of slaves to be marginalized, however.

Let all Americans take a few years, a few decades if necessary, and just get out of each other's faces. Let the healing begin with a cooling-off period. Practice mutual respect and common decency, but give the social engineering a rest. Try equality instead of racial integration. Let us choose our neighbors, schools, jobs, and friends on the basis of personal choice, desegregation, equal education, and individual character.

If we repudiate racism and dominance, our ancestors will thank us for allowing their souls to finally rest in peace.

CHAPTER 6

NEGATIVE IMAGES, NEGATIVE RESULTS

It's not what you call me, it's what I answer to.
—AFRICAN PROVERB

Like lethal radiation that seeps up through a fissure in the earth, the poison of three hundred years of slavery, Jim Crow laws, welfare plantations, self-serving leadership, and racial stereotyping surfaced in a sixth-grade public school class in Montgomery County, Maryland. The enduring effects of racism and self-victimization revealed themselves when Black students in the class were asked to speak on their impressions of Black people.

"Everybody knows that black people are bad. That's the way we are," one Black student began. Of the twenty-nine Black students present, twenty-four agreed with the speaker that they were inferior, according to a *Washington Post* article appropriately entitled "Stereotype Within."[1] Most believed that this inferiority was inherited. No matter what happened during their upbringing, it would not change that fact. And education was absolutely useless in their world, they believed.

The students had such an unfavorable opinion of themselves that they accepted the stereotyped image. The *Washington Post* article outlined the tragic results of negative self-image and stereotypes among the young Black students. In general, they believed that:

• Blacks are poor and stay poor because they are dumber than Whites and Asians.

• Black kids who do their schoolwork and behave must want to be White. White kids who do poorly or dress cool want to be Black. Hispanic kids want to be Black because they aren't smart like Whites.

• Black people don't like to work hard.

• Blacks don't need to work hard because it won't matter in the end.

• Black people have to be bad so they can fight and defend themselves from other Blacks.

• Blacks see their badness as natural.

• Black men make women pregnant and leave.

• Black boys expect to die young and unnaturally.

• White people are smart and have money.

• Asians are smart and make money.

• Asians don't like Blacks or Hispanics.

• Hispanics are more like Blacks than Whites. They can't be White, so they try to be Black.[2]

In their striking rejection of their own culture, these young people were not defining themselves. They were defining their environment—including the racists, rappers, politicians, celebrities, entertainers, and race leaders who shape it. Most of all their parents have likely instilled in these children these self-defeating attitudes. This self-victimization has become a cultural legacy for many Blacks and the greatest barrier to Black progress, racism notwithstanding.

Stereotypes are deadly even within the Black community, where many young people have learned to live down to expectations. A Black man told me that my socioeconomic status could only be explained because I was "chosen" for success by White people. The stereotype that he has internalized of himself makes him believe that external circumstances (White people) determine his worth. Conversely, failure for most Blacks is the result of not being chosen by Whites. And, of course, Whites will only allow a handful of Blacks to do well, the self-confirmed victim argued. Symbols and stereotypes can be devastating.

The Black population in America has historically been marginalized by the manipulation of images in the media. Both Whites and Blacks have been conditioned to believe that Whites are superior and Blacks inferior. This has been accomplished by creating negative symbols that define Blacks: poverty, crime, violence, unemployment, irresponsibility. In fact, Blacks are defined by their problems. This is the tragedy of Blacks in America—that the image is the message and the media is the image.

Blacks have been dehumanized as coons, brutes, mammies, shuffling darkies, predators, and sex fiends. Today's Black gangsta culture embraces those stereotypes and sings their praises in rap form, glorifying sociopathic behavior. How would you like to be a Black child growing up in that environment? Under this psychological barrage, nearly all

Blacks wind up psychologically damaged to some extent by corrupted images and low self-esteem.

Media Miseducation

Dr. Carter G. Woodson explained the origins and long-term effects of low self-esteem in his farsighted and classic 1933 book *The Miseducation of the Negro.* "When you control a man's thinking you do not have to worry about his actions. You do not have to tell him not to stand here or to go yonder. He will find his proper place and will stay in it. You do not need to send him to the back door. He will go without being told. In fact, if there is no back door, he will cut one for his special benefit. His education makes it necessary."[3]

Although there are a few exceptions, a young Black person will likely seek his or her "proper place" according to the stereotype. Many who are part of the underclass in impoverished and gang-controlled areas will join the gang-bangers on the streets and, it is likely, go to an early grave. Even a few of those lucky ones who escape and join the middle class are most likely to follow the stereotypical path. They become civil rights or church leaders, entertainers, athletes, instead of pioneering new fields or following their individual talents and instincts. We have not been conditioned as a people to assume that our "rightful place" might be to be the President of the United States, the owner of a media conglomerate, or even a marine biologist or nuclear engineer.

In the minds of liberal racists like Francis Lawrence, the president of Rutgers University, SAT tests are unfair to Black students because they lack the "genetic, hereditary background" to make high scores. As a champion of liberal racism's belief in Black inferiority, Lawrence has insisted on race norming and an assortment of liberal panaceas throughout his academic career. With liberal "friends" like this, who needs conservative racists?

Conservative racists block the door to opportunity. Liberal racists stifle Black self-reliance. Both foster their messages in images. The master manipulators of opinion, whether in politics, show business, or advertising, understand that the image is the message. And many of them exploit this knowledge for nefarious and racist purposes as well as for profit.

For example, the motion picture industry in the early twentieth century acted out the expectations and perceptions of a racist society by projecting the old Southern rural stereotypes of Black life: White actors, daubed with burnt cork, splattered America's image of Blacks on the screen. And this is what America came to see Blacks as—dimwitted, shuffling darkies with a childlike mentality and a submissive nature.

Lingering Images

Through the decades, the Black image has been ridiculed and vilified in popular American culture and exploited for the commercial benefit of Whites. The Black stereotype has been reinforced and spread around the world in films and on radio and television. The Japanese, in particular, are often American-culture clones. They have been influenced by Hollywood's racist depictions to the point that racist images of Blacks are widely accepted in Japan. And the social and economic impact of this global racism can be enormous.

A 1989 study by Robert E. Cole and Donald R. Deskins, Jr., concluded that Japanese auto manufacturers and suppliers in the United States intentionally locate their plants in areas with relatively few Blacks and "had a lower percent of Black employees."[4] The authors say their findings "suggest that the Japanese have 'a taste for discrimination.' "[5] Many "Japanese companies . . . specifically asked to stay away from areas with high minority populations,"[6] they note. Some Japanese ask for community profiles broken down by ethnic background in order to find "a high German content," and a Japan External Trade Organization publication identifies California as a good site selection because of the large Asian population.[7]

Another study on Asian investments in the United States found that the American stereotype of Blacks as "streetwise, rioting, stealing or drugged" is the most common depiction of Blacks in novels read by the Japanese overseas businessman.[8] Some Japanese make much of their negative feelings toward Blacks, but rarely mention that the GM-Toyota joint venture in Fremont, California, which has the highest percentage of Blacks (23 percent) and Hispanics (28 percent) in any of their plants, is a model for quality and productivity.[9]

Self-Destructive Self-images

Hateful images and ideas crush "the spark of genius in the Negro by making him feel that his race does not amount to much and never will measure up to the standards of other people,"[10] according to Dr. Woodson, the father of Black History Month.

This chronic lack of self-esteem among Blacks is most evident among the young, but it manifests itself also in the behavior of adult Blacks—even those who put themselves forward as role models and leaders. Consider the impact on young Black minds when the following widely quoted statement was made by Jesse Jackson, the man considered by many Blacks to be their premier leader: "I hate to admit it, but I have reached a stage in my life that if I am walking down a dark street late at night and I see that a person behind me is White, I subconsciously feel relieved."[11]

Another member of the Black Talented Tenth, professional Marxist Cornel West, who lives regally on a six-figure income while encouraging Black college students to discard capitalism, has offered this negative dismissal of poor Blacks: ". . . without jobs and (economic) incentives to be productive citizens, the black poor become even more prone to criminality, drugs and alcoholism."[12]

When Black intellectuals write off all Black poor as being naturally "prone" to crime, drugs, and alcohol, what are poor young Blacks to think? Are they simply doomed? Do they never stand a chance of enjoying the sort of life led by the effete elitist Mr. West?

These statements by West and Jackson have a disheartening similarity to the defeatism and self-loathing that permeates the statements made by the Black sixth-graders in the Montgomery County classroom. These are Black opinions built on negative racial stereotypes. As happens more often than we sometimes admit, the attitudes and behavior of some "role models" project racial inferiority and low self-esteem rather than racial pride. By making such public statements, Jackson and West exacerbate the low self-esteem of some Blacks and the racial neurosis of some Whites—such as liberal racists like Lawrence at Rutgers.

Undoubtedly, the remarks by Jackson gave comfort to Whites concerned about the deterioration of urban Black life because they symbolically signaled a long-overdue admission of Black responsibility. But racist stereotypes feed the neurotic predisposition of some Whites to believe that crime committed by Blacks is racially and culturally determined.

I have never been comforted by the presence of a White person walking behind me on the streets—because it is not the race of the person that brings safety or harm, but the level of criminal activity of a neighborhood. Being followed closely by anyone in a high-crime area is cause for raw nerves. If you're in an all-Black neighborhood, the chances of the criminal being Black are excellent, just as they would be for the criminal being White in an all-White area.

And, as a product of poverty who has risen to some level of accomplishment, I resent West's statement as well. The issue is poverty, drugs, and crime—not Black, White, Asian, Hispanic, or Native American. The Black predicament of negative imagery is created by the stereotypical manipulation of racial symbols and images. The result is that complex social and economic phenomena are blamed on Blacks, who find themselves marginalized and scapegoated—dangerous for any ethnic minority group. As I said earlier, Blacks are defined by their problems.

As Jews discovered in Nazi Germany, negative imagery can lead to very real problems. Once a group is marginalized, it is easily scapegoated. Blacks have been reminded of this frequently in recent times. In 1989, Charles Stuart, a yuppie retailer, convinced most of the world that an armed Black gunman had shot him and killed his pregnant wife as they left a childbirth class at a Boston hospital. Blacks were hauled in for questioning, until a family member eventually testified that Stuart had wounded himself and murdered his wife. More recently, in November 1994, Susan Smith, a twenty-three-year-old White mother from Union, South Carolina, told police that a Black man had pulled her from her car and driven off with her two sons, aged three and fourteen months. A few days later, after an intensive manhunt in which dozens of Blacks were interrogated, Smith finally confessed to killing her own children. Whites were shocked, but Blacks were, to a large degree, relieved that the onus was off them—at least in that instance.

Promoting the Wrong Image

Black leaders who promote destructive images of Blacks to hype themselves do their racial community no favors.

After seeing a Khalid Muhammad performance on the nightly news, I can imagine at least half of White America under the bed trembling, waiting for the Black Avengers to break down the door. I can also picture the other half of White America in militia khakis waiting with bated breath and AK47s to send Muhammad and the first Blacks they see back to Allah in a hail of bursts from automatic weapons.

Images of violence provoke reaction. Every Black fanatic understands that while it is safe for Blacks to threaten and murder Black opponents, the route to instant fame is to threaten or to seem to threaten to kill White people. *News-*

week promptly rewarded a nonrapping "rapsta"—Sister Soul-jah—with the magazine's cover. Khalid Abdul Muhammad has been making the same vile assertions in the Black community for over a decade as Farrakhan's national spokesman. But he couldn't draw flies in the Black community until his infamous speech to only about a hundred people at Kean College on November 29, 1993.

After that virulently scornful diatribe was advertised in the *New York Times* by the Anti-Defamation League and spotlighted by the mainstream media, the self-described "truth terrorist" began to draw thousands, Black and White, to his performances, for which he reportedly demands $10,000. Without the media push and ADL's fundraising promotion, Muhammad would have remained in obscurity, preaching racial hate to those already full of bile.

Far too often, the Black community is more concerned with the image than with the reality. And in the long run, it hurts the community badly, because a corrupt leadership is worse than no leadership at all.

The advancement of Blacks on all fronts—especially in ownership and on-air presence in the mass media—had to figure into the thinking of the White teacher back in Montgomery County, Maryland, when he wondered how these White racist stereotypes persist. How does this low self-esteem/self-hatred syndrome maintain such force in the 1990s? We know about the role of racism, but the answer may also come in part from recognizing that we are not weighing the impact of another factor: role models—or negative images that cause deadly results.

Media and schools are the traditional and easy scapegoats because their failures are so conspicuous. Black leaders blame the media for the distorted values of Black criminals. Jackson, for example, likes to point out that by the age of fifteen, the average youngster has seen eighteen thousand hours of television and 500,000 fictional murders, but has spent only eleven thousand hours in school and three thousand hours in religious services. Those are interesting and appalling statistics, but it is equally appalling that Jackson would suggest that all Blacks are criminals and then accept no blame or responsiblity for allowing these subliminal messages to flourish.

Reality Versus Image

Blacks remain the stereotype for impoverished and dependent racial groups, but in reality, most Blacks are not poor. The American Black community will earn more money this year than many countries in the world. Nearly 20 percent of Black households earn over $50,000 a year. Poor Blacks are losing their small share of the country's wealth, but so are poor Whites. Families earning less than $15,000 increased from 34.6 percent to 37 percent of the Black population between 1970 and 1990.[13] However, the number of Black middle- and upper-income families rose at a faster rate. Black families with incomes between $30,000 and $50,000 increased from 13.9 percent to 15 percent of the Black population.[14] And, astonishingly, Black families earning over $50,000 rose from 10.2 percent in 1970 to 16 percent of the Black population in 1992—a faster rate of increase than among White families. The percentage of Black professionals and managers increased from 10 percent to 16.8 percent, and Blacks between twenty-five and twenty-nine years old who graduated from college more than doubled during the same period.[15]

Black Greeks

All too often, when Blacks find something to take pride in, or to focus on as a source of pride, others attempt to tear away at it. I am thinking in particular of Black fraternities and sororities, which have become an increasingly important social and career network. I know of lifelong friendships and also of entire businesses built through relationships that began in these organizations. I remember my own intense pride and the honor I felt when I was inducted into Alpha Phi Alpha at Wayne State University in Detroit. The fraternity, which also claimed the Rev. Martin Luther King, Jr., as a member, is a social organization, but its overriding emphasis is on moral virtues—"manly deeds, scholarship and love of all mankind."

And yet, there are those Blacks and Whites who attempt to discredit these highly beneficial organizations by attacking the concept of "Black Greeks" and criticizing members for trying to be like White Greeks. Well, I have news for those critics. No, I have a little history for them. Ancient Greece was a culturally diverse, pluralistic nation much like the modern United States.

The Pelasgians: Black Greeks

Pelasgian, or Pelasgoi, was the name "given to unassimilated native Greeks,"[16] writes Martin Bernal in *Black Athena*.

Some researchers believe that these earliest inhabitants of Greece were a dark-skinned, brown-complexioned combination of groups: Phoenician Canaanites from Asia Minor and Palestine, and Egyptians and East Africans. The sixth-century poet Asios of Samos is quoted as writing: "And black earth produced the god-equalling Pelasgos."[17] It was only after the Aryans invaded Greece around 600 B.C. and before 1850 that the "Greek origins was transformed" to Aryan myth,[18] according to Bernal.

These historical findings make me proud to celebrate the millions of Black Greeks and their fraternities and sororities. I am proud to be descended from great Pelasgian people like Homer, Socrates, Euclid, Ulysses, Achilles, and Hercules. And I am very proud to be a Black Greek.

Black Families Thrive

The *New York Times* recently discovered that Black families in the middle-class borough of Queens surpassed the median income of White households during the 1980s, when the number of Black families earning over $50,000 doubled.[19] Two factors were cited in explaining the phenomenon of unstereotypical Black success—a strong family structure and hard work. A closer look at economic progress during the 1980s—the Reagan years—may shatter another stereotype: that of universal Black deprivation. The overall impact during the Reagan years was on class, not race. Because of the demand for more technical skills in an information economy, the wealthy and upper-middle class grew faster than the middle class in the Black community. But do not look for that story in the media. The image of Reagan the Racist holds the liberal army together by unifying Blacks against an objectified conservative evil.

Among Black liberals, it is considered impolite to suggest that President Reagan may not have been anti-Black, or that if he was, it did not stop Black progress for the upwardly mobile. Black-owned businesses increased by one third during the 1980s. Overall, the poor, Black and White, did not progress equally because of a growing high-tech-knowledge sector and a diminishing manufacturing base. The BUM's grip on the Black community needs Reagan the Racist to make the rest of the equation work. But the truth tarnishes the myth.

In fact, I was able to solicit then President Reagan to change the direction of the government from closing the nation's Black colleges, which the liberals under Jack Greenberg at the NAACP Legal Defense and Education Fund (not a part of the NAACP) were doing. In response to my pleas, Reagan even increased the budgets and aimed several programs specifically at Black colleges. This is strange behavior for a racist. Insensitive to Black culture and intolerant of Black socialists, he was. For his being racist, other than his support for the Bob Jones University segregation policy, I find little other evidence.

Between 1970 and 1990, America moved closer to becoming a nation of haves and have-nots; it was already a country divided along rigid racial lines. Technological displacement had as much impact as Reagan's policies, if not more. Economic class lines have more to do with success or failure than being Black or White does. While we're bursting bubbles, consider that the rioters who were arrested following the 1992 Rodney King trial were 36 percent Black as compared to 51 percent Hispanic, according to follow-up studies. The media image of the rioters was almost 100 percent Black. Have you read that inner-city Black youths are leading a general decline in illicit drug, alcohol, and tobacco usage? A study by Partnership for a Drug-Free America found that only 0.8 percent of Black high school seniors use cocaine (it was 5.8 percent in 1986) as compared to 3.1 percent of their White peers (it was 13.5 percent in 1986). The spread is even larger for smoking (32 percent for Whites, 8.2 percent for Blacks) and heavy drinkers (32 percent for Whites, 11 percent for Blacks).[20] By 1993, according to the *Journal of the National Cancer Institute*, just 4.4 percent of Black teens smoked.

Why are Black youth making healthy choices? Thank the Black churches, mosques, and synagogues, and most of all, thank the Black parents who teach wholesome values. What? Responsible Black parents? That image doesn't exist outside of Cosby's world. The media do not promote it because it does not meet the sterotype, and it does not sell soapsuds. The media put a Black face on drug abuse among the young, no matter what the facts say. In Great Britain, Africans are twice as well educated as Whites and twice as likely to become professionals.[21] But this story has been ignored by the U.S. media because it would conflict with the image of downtrodden, illiterate Blacks in America who suffer from a chronic case of *The Bell Curve*.

The media are not alone in promoting lies and false images. The history of false prophets is four thousand years old, and some Black leaders have helped write it. Today the BUM's leaders sometimes find it necessary to sell "wolf tickets" and play the race card when reality intrudes on image-making or when they need to stay out of jail. When caught with their pants down, or their pockets full, the Black demagogues respond by attacking the messenger. Deposed

NAACP leader Benjamin Chavis tried to make the media and Jews the issue and racism their motive when he was caught using the NAACP's funds to pay off a woman who charged him in a lawsuit with sexual harassment. What Chavis failed to understand is that his conduct cannot be hidden or written off as the creation of racists. His grandstanding stewardship as head of the NAACP was inept and forcefully intolerant of Black people who disagreed with his Marxist philosophy. Among Chavis's plethora of self-inflicted problems, his advocacy of obscene rap music forced the NAACP's board of directors to repudiate his endorsement and condemn rap lyrics that uphold violence and degrade women.

Bad Rap Images

If anyone should criticize the gangsta rap filth, it should be Black leaders, because it is Black teens who created the phenomenon. To the defenders of gangsta rap, any link with the past "justifies any excess,"[22] Martha Bayles writes in *Hole in Our Soul*. She cites Harvard's Black scholar Henry Louis Gates and, of course, Chavis, both of whom endorsed gangsta rap as a Black art form. Such uninformed endorsements from prominent Black leaders make misogyny, violence, and vulgarity acceptable forms of behavior and encourage the degradation of women and the exploitation of young minds. The money machines that crank out rap music are awesome. Much of rap is motivational and poetic, but as in any other genre, the truly rotten stuff stinks up the entire scene.

Music experts and critics such as Nelson George describe rap as a "social triumph."[23] Even a middle-aged Black man like me can accept that there are cultural merits in music that comes out of shared experience. It's when Niggaz with Attitudes, Ice-T's "Evil Dick," and 2 Live Crew's "As Nasty as They Wanna Be," among others, come into hearing range that I reach for the off-button or doorknob.

I don't think you have to be middle-aged or a psychologist to know that violence, vulgarity, and hating women are symptoms of mental illness. I once developed a course called "Psychology of the Ghetto" that I taught at the old Federal City College in Washington, D.C., and later at Howard University. It applied the disciplines of sociology and psychoanalytic theory to the Black experience—the pathology of the Black experience, to be more precise. The phenomenon of gangsta rap fits that genre and cries out for a nontraditional explanation. Gangsta rap songs are packed with misogynistic boasting of oversized sexual features (exaggeration is a sure clue that the advertisement is false) and disgusting accounts of "bitches" and "ho's" being raped, abused, and sodomized. These violent themes reveal more about their creators and performers and our afflicted society as a whole than about the women victims they intend to defame.

One gangsta rapper told a convention of the National Association of Black Journalists that he called women "bitches" and "ho's" because in his mind that's what they were. A large part of the audience had enough self-esteem to walk out. But Black Los Angeles Congresswoman Maxine Waters calls gangstas such as that foul-mouthed lout "poets" and "our children." She credits them with having "invented a new art form" in response to this awful place called America[24]—the place that Cubans and Haitians are fighting off sharks to get into. Therefore, she and others suggest, society should refrain from "a course of censorship."[25] The rest of us, she seems to suggest, should climb down in the gutter with the gangsta rappers, ignoring the old saying "He who lies down with dogs shall rise with fleas."

If Blacks are to survive racism, they must first survive their own leadership. Many Talented Tenth elitists do acknowledge the moral decline that this gutter filth propagates, as well as the political and social barriers it builds to block Black progress by reinforcing the old stereotypes.

C. DeLores Tucker of the Political Congress of Black Women and the Reverend Calvin Butts, a Harlem icon, recognize the link between these negative images and the deep racial hostility that exists in some segments of the White community. In a 1993 survey conducted for the National Science Foundation, 51 percent of White conservatives and 45 percent of White liberals felt that "blacks are aggressive or violent." I feel the same way about gangsta rappers. Unfortunately, most Whites don't differentiate between this small group of gangsta rappers, other rappers, and the rest of the Black population. It is fair to say that the vulgar, hateful, sexist, often racist lyrics are the result of low self-esteem—not of material poverty, not of racism, and not of the absence of an adult male in the home. Many single parents raise well-adjusted children who do not become gangstas, but these Black family success stories are rarely portrayed in the media or acknowledged by Black leaders. I come from such a home and so do many of the successful Blacks whom I know personally.

Self-hate is a product of poor self-esteem, which results from a background lacking in love and moral virtues. This can happen in any household, rich or poor. Witness Jeffrey Dahmer, one of history's worst killers, a cannibal who came out of a middle-class White two-parent household, as did Laurie Dann, who shot up a suburban Chicago school.

Honest, hardworking Blacks have enough to deal with in a society increasingly controlled by demagogues and sociopaths. Negative images bombarding them from all sides, and from within, only make it more difficult to stay on the righteous and moral path. Black and White leaders have a moral obligation to society to strike down, not perpetuate, negative and racist images and see that the truth, not the stereotype, endures.

CHAPTER 7

"AIDS" OR "DAIDS"?

Suppose HIV doesn't necessarily equal Aids. Then we will have witnessed the biggest medical and scientific blunder of this century.
—NEVILLE HODGKINSON in the *Sunday Times* (London), April 26, 1992

The last time I saw my friend Arthur Ashe alive was on October 19, 1992. Ashe, a former Wimbledon tennis champion who contracted HIV through a blood transfusion, appeared with nutrition expert Gary Null as my guest for a taping of a segment of *Tony Brown's Journal* entitled "The AIDS Cover-up."

Arthur's last words to me on that program were in defense of AZT, the chemical therapy for HIV infections and "AIDS." He obviously wanted to believe that the drug prolonged life, but he was plagued by doubt.[1]

I have no such doubts about AZT. I think it killed my

friend sixteen months after we appeared on that show together. And I think AZT has killed untold others.

The toxoplasmosis disease Ashe contracted in 1988, after surgery in 1983, disappeared, but because he was HIV-antibody-positive, his case was still regarded as "AIDS" according to the standards of the Centers for Disease Control and Prevention in Atlanta, Georgia, commonly known as the CDC.

According to the medical establishment's standard treatment, AZT was recommended as a treatment for Ashe. Every day the great athlete's immune system was assaulted by an initially very large dosage of AZT. Later, when he reflected on the effects of that chemotherapy, Ashe said, "I refuse to dwell on how much damage I may have done to myself taking the higher dosage."

In addition to highly toxic AZT treatments, Ashe was taking drugs daily for heart trouble, high blood pressure, and high cholesterol. There were also antibiotics administered to stop opportunistic infections. He took "cleocin to fight further toxoplasmosis, nystatin to slow down yeast infections, and pentamidine to stave off Pneumocystis pneumonia. Two other drugs were prescribed against possible brain seizures,"[2] according to Peter Duesberg and Bryan Ellison in their book *Inventing the AIDS Virus*.

Of the thirty pills Arthur took each day, only a few were vitamins. Understandably, this great sports figure began to waste away. Ashe was unable to break away from the psychological pressure of his doctors and their chemical regimen, and so his "poisoned immune system could no longer fight off" death,[3] Ellison concluded.

On my PBS show and in other interviews prior to his death on February 6, 1993, Ashe expressed his "confusion" over the relationship between HIV and "AIDS," and on whether AZT is an appropriate treatment for those who are HIV-positive. In the end, the aristocratic Ashe accepted the standard treatment for "AIDS" because he felt it was his only hope. He gave the benefit of the doubt to modern "AIDS" medicine and the devastating chemotherapy.[4]

I believe Arthur Ashe could have lived much longer. Perhaps even to a ripe old age. It can be argued that the "AIDS" treatment, not "AIDS," killed him.

In the preceding chapters, I have outlined some of my concerns for the future of this country. In this chapter and the next, I want to address my fear for the lives of all of mankind, a fear of death from what I have come to think of as the biological equivalent of the devastating nuclear accident in Chernobyl. This, you may recall, is the city in Ukraine where the world's worst nuclear reactor accident occurred in 1986. When the cooling systems of one of the plant's reactors failed, the core overheated, resulting in an explosion and fire that took eight thousand lives. The radioactive fallout contaminated large areas of Eastern Europe and Scandinavia, and its impact will be felt for generations to come.

I believe there is a 100 percent chance of worldwide catastrophe greater than Chernobyl. We have not been told the truth about this biological threat. And what we don't know may kill us. After years of examining this issue and after conducting dozens of interviews with scientists and medical experts on my PBS television series, I have concluded that "AIDS" is a scientifically dishonest construct, and, as exemplified in the death of my friend Arthur Ashe, the lies that we have been told make it all the more deadly.

You will notice that throughout this chapter and the next, I do not mention "AIDS" without enclosing it in quotation marks. I do that to make a point: "AIDS" is still an unknown quantity. A growing number of people do not accept the theory that "AIDS" is caused solely by the human immunodeficiency virus (HIV).

The standard theory of "AIDS" and its origins has been accepted as gospel, but I am among those who think it is a lie. The medical establishment would have us believe the bizarre theory that "AIDS" originated when a monkey bit an African on the butt. The theory goes that the bitten African then went home and engaged in debauched sex and seeded the human species with a modern plague.

This is absurd. It is an obvious attempt to blame this alleged disease on Africans in particular—and Blacks in general. This is yet another poorly masked attempt to marginalize Blacks. And so I have made it a mission to disprove the lies and to search for the truth. Over the last decade, I have written and spoken on the topic across the country. I was the keynote speaker at the Delaware HIV/AIDS Awareness Conference in the fall of 1992, and the *Delaware State News* reported that Tony Brown "startled many in the audience when he declared that he does not believe that HIV causes AIDS."[5] The newspaper noted also that my opinions on the topic of "AIDS" dissented from commonly held opinion and therefore "may not be welcomed by many."[6]

That certainly is true. My opinions differ greatly from the propaganda put out by the medical establishment. Most of those in the "AIDS" industry do not welcome challenges to their doctrine. But I have challenged them and will continue to do so. I have produced more national television programming on the topic of "AIDS" than any other source, and the programs have brought a greater viewer response than any others I have done.

Victims of "AIDS," in particular, have responded to my challenges to the medical establishment. Many of them realize that they have been victimized as much by lies as by some insidious viral disease. None of the medical establishment's pronouncements on "AIDS" should go unquestioned. Not with the stakes so high, and not when lies have been spread from the very first.

The potentially dangerous story of the disease's African origins arose in 1985 with a veterinarian from Harvard named Dr. Myron "Max" Essex, a specialist in animal retroviruses. He was an associate of Dr. Don Francis, then a virologist at the CDC, and of Dr. Robert Gallo, a controversial virologist at the National Cancer Institute. Gallo has become the point man in our country's "AIDS" research and one of the chief financial benefactors of the hypothesis that HIV equals "AIDS."

These men and other virus hunters were veterans of the failed War on Cancer that began in 1971. They and others spent decades, not to mention $23 billion, trying to prove that a virus causes cancer, but these great virus hunters failed. The rate of cancer has increased every year, and it is projected to be the nation's biggest killer within the next five years. And now these same scientists are chasing a virus that they claim causes "AIDS." They have yet to prove that a virus causes it, but the "AIDS" industry is already a trillion-dollar monster.

It was the CDC's Francis who first claimed that "AIDS" is caused by a virus. He then sold Essex on the virus theory, and Essex sold it to Gallo, the man with enough clout to sell the theory to politicians at the U.S. Department of Health and Human Services and the virus hunters at the National Institutes of Health. French scientist Dr. Luc Montagnier of the Pasteur Institute discovered the so-called "AIDS" virus, which Gallo claims to have found independently a year later in the United States. Virtually overnight, these former cancer research scientists became internationally acclaimed "AIDS" experts. Their failed search for a cure for cancer was all but forgotten in the glory of their new mission to save humankind. After all, "AIDS" was hyped immediately as potentially the most deadly virus ever known.

Like many who have made a serious study of viruses and the harrowing world of run-amok biomedical research, I fear there are even deadlier microorganisms to be unleashed on the world; I call these "Chernobyl II" microbes, and I will deal with Ebola and those yet unnamed in the next chapter.

Both of these potentially catastrophic biological threats have come from medical research laboratories, where, we have just begun to learn, the blind pursuit of scientific glory and financial gain overrides any sense of responsibility for human life. Biomedical research has built a disease-making machine and endangered humanity as a result.

"AIDS": The Cash Cow

The United States spends more on health care than any other nation in the world: 12 percent of the gross national product, or $650 billion in 1991.[7] We have already spent $35 billion for "AIDS" research, yet 270,000 people have died from a treatment that has yet to save a single life. By the year 2000, the cancer and "AIDS" establishments of the government-academic-medical-industrial complex will spend $1.7 trillion as the total health care bill for Americans.[8] The taxpayer-funded health care explosion is now subsidized largely by the "AIDS" scam—research medicine's new cash cow.

Incredible amounts of money are being made in the booming "AIDS" industry. The two men who claim discovery of the HIV virus, Montagnier and Gallo, earn patent royalties for the so-called "AIDS" test; in reality, it is an HIV-antibody test. This test, which many believe is unreliable, cost more than $1.2 billion to develop. You know that the $1.2 billion didn't come out of the pockets of Montagnier or Gallo.

Viruses are discovered nearly every day, but few scientists have at their disposal the public relations and marketing force of the federal government to promote their discovery, or to promote their test as the only way to detect what was quickly billed as the era's most deadly epidemic. People who believe they are HIV-positive are being scared into mass hysteria and suicide by the "AIDS" industry's self-serving medical disinformation. HIV patients fear an inevitable and horrible death when, in fact, they may not be in imminent danger—unless they take the chemical treatment prescribed for "AIDS" patients.

The greatest threat to life in the United States and the rest of the world is not the grouping of approximately thirty diseases (including one non-disease) called "AIDS"; it is the $25 billion "AIDS" industry composed of government, academic, medical, and industrial interests in the United States. This medical cabal is controlled by an ambitious group of researchers within the National Institutes of Health. These researchers and their cohorts have usurped the U.S. government's delegated authority to preserve life and good health among the American people.

What these virus hunters are doing is not quite geno-cide; it is triage for fame and fortune. Their "treatment" for "AIDS" results in death and dying. We must stop these tax-supported angels of death in our government's scientific establishment.

What Is the "AIDS" Epidemic?

It has become abundantly clear, in spite of a great campaign of disinformation and reprehensible scare tactics, that "AIDS" does not attack the general population. After fifteen years, it remains almost exclusively confined to IV-drug users (about 32 percent of the cases) and a subset of male homosexuals which accounted for about 60 percent of the total 140,428 AIDS cases in the United States in 1991.[9] The total number of homosexuals who have had "AIDS" since it was discovered in 1981 was 217,012 as of December 1993.[10]

The key to the debate on the so-called "AIDS" epidemic is understanding how many homosexual men *do not* have "AIDS." *Nearly 3 percent of the eighty million adolescent and adult American men are homosexuals—or approximately 2.4 million. Therefore, of 2.4 million homosexual men in the United States, 93 percent do not have "AIDS"; 7 percent do. On the other hand, if the larger figure of 5.3 percent of men who have had homosexual relations at least once since puberty (the estimate for women is 3.5 percent) is considered to be the homosexual population, only 5 percent of homosexual men have "AIDS," and 95 percent do not.*[11] If homosexual sexual practices per se were the cause of "AIDS," how could 93 percent or 95 percent of homosexual men not have "AIDS"?

And have you ever heard of even a single case in which a homosexual outside of the risk group infected another non-risk-group homosexual with HIV? Or transmitted some other microorganism that led to an "AIDS"-related disease? I haven't.

As the noncorrelation between homosexual male sex and "AIDS" is examined, it becomes apparent that the incidence of "AIDS" correlates only with drug abuse. About 95 percent of those who contract "AIDS" have a history of drug use—according to Dr. Robert E. Willner in his book *Deadly Deception*. Willner quotes studies that claim it takes from "500 to 1,000 unprotected sexual encounters to transmit" HIV, the magic-bullet virus blamed for the thirty or so separate diseases that the "AIDS" establishment has designated as "AIDS."[12]

Even the diehards in the "AIDS" establishment have been forced to admit the strong correlation between "AIDS" and drugs. The *New York Times* reported in February 1995 on a CDC study that found three-quarters of the forty thousand new HIV infections reported in 1994 were among drug users.[13] "AIDS" is not the leading killer of young adults; drugs are. And I count among those killer drugs the very ones that are being used to treat the diseases we know as "AIDS."

The odds of a healthy non-drug-using heterosexual getting "AIDS" are the same as for getting hit by lightning. And from a population of 255 million Americans, only 140,428 were living with "AIDS" as of 1994.[14]

John Lauritsen and Hank Wilson, in their book *Death Rush*, accuse the CDC of fraud: "The effect of the CDC's statistical trickery is to underreport IV-drug users as an AIDS group by at least 50 percent; the effect is to construe AIDS as a venereal disease, rather than a drug-induced condition."[15]

Bad Medicine

In this chapter, I will explain to you how I have come to believe that the U.S. government has lied to the American public and the world about the true nature of "AIDS." And I will offer you my theory that the National Institutes of Health (NIH) and its "AIDS" complex—including the two most nefarious units of propaganda, the Centers for Disease Control and Prevention (CDC) and the Food and Drug Administration (FDA)—are scamming the American taxpayer out of billions of dollars each year.

The "AIDS" industry is the heir apparent of the nearly forgotten cancer industry. For all of the millions that have been poured into it, the "AIDS" industry has produced no new science to prove its theory that the HIV virus causes "AIDS." It has only succeeded in sowing the seeds of fear and hatred between various population segments of this country. It has branded those with HIV antibodies as pariahs. It has unjustly marginalized innocent people, and even killed them.

Our government agencies charged with protecting the nation's health are instead busting the health care budget to prove their unprovable theory that a virus called "HIV" causes "AIDS." And they profit greatly from it.

The bureaucrats generate billions of dollars from royalty payments for patenting the so-called "AIDS" test. They have approved a "treatment" (AZT) that thirty years ago was considered so deadly that it was taken off the market. And they have been selling an "AIDS test" that renders false positives an estimated 20 to 50 percent of the time in the United States.

In Africa, Zairian researchers have shown that certain bacterial illnesses may cause as much as a 70 percent rate of false positives on both standard HIV antibody tests, according to a report by journalist Neenyah Ostrom of the *New York Native*.[16] Pregnancy can also trigger false positives. In addition, reporter Laurie Garrett notes, in her well-researched book *The Coming Plague*, that the reported HIV-positive rates of 50 to 90 percent among people living below Africa's Sahara could be grossly off, because nearly everyone in that region carries some malarial parasites in the blood, which may affect HIV test results.[17]

All of this chicanery began with the unproven assertion that the HIV virus causes "AIDS." This hypothesis was formed without traditional standards of scientific verification. There are no scientific reference papers proving the HIV-"AIDS" hypothesis, no animal models in which test subjects have been injected with HIV and contracted "AIDS." The hypothesis even fails Koch's Postulates, a set of four scientific rules to determine whether a specific microorganism causes a specific disease.

The U.S. government is bankrupting itself to fight an "AIDS" *epidemic* that is actually restricted to very specific population subgroups of drug abusers and others with unhealthy lifestyles (along with those they infect) that decimate their immune systems. This is not indicative of a true epidemic by any medical standard.

This so-called epidemic is being kept alive only by artificial means. The number of "AIDS" cases has grown in large part because of the ever-expanding official definition of "AIDS." That definition has been changed four times in eleven years to include more and more illnesses—most of which have been around for a long time. Cervical cancer, for example, is now included in the list of diseases under the "AIDS" rubric if HIV is present.

In 1985 the first change in the *definition* of "AIDS" resulted in a 4 percent increase in the diagnoses, or two thousand new cases from more old diseases. They were suddenly brought under the "AIDS" umbrella. Another redefinition in 1987 added 30 percent—or nearly ten thousand—more cases. In 1993, women with cervical cancer and people who had only low CD-4T cell counts but were not sick were classified as "AIDS" patients. CDC then promoted an "AIDS" epidemic among women and added between fifty thousand and two hundred thousand new "AIDS" victims.[18]

According to Tom Bethell in *The American Spectator*, if the 1987 definition of "AIDS" were in place today, the cumulative total of "AIDS" cases would be reduced by more than half, and the number living with "AIDS" might be no more than twenty thousand.[19] And yet, the White House AIDS Policy Office reports that approximately two hundred thousand Americans are living with "AIDS"—based on today's suspect definition of it. Now you know why Mark Twain said there are three types of lies: lies, damned lies, and statistics.

And isn't it peculiar that the incubation period of "AIDS" (the length of time between when you contract the virus and when you begin to exhibit symptoms of an "AIDS"-related disease) keeps lengthening? It originally was twelve to eighteen months, then went to five years, then seven, then ten, then fifteen, and gradually to a point nearing the duration of a normal lifetime. At this rate, we will one day have healthy ninety-nine-year-olds who have carried the "AIDS virus" since the age of twenty.

My attitude toward "AIDS" is simply an amplification of the position of a growing number of senior scientists and medical doctors. My sources include the most prominent dissenting voice of all, Dr. Peter Duesberg. Perhaps the top authority in the world on viruses and retroviruses, he is a member of the prestigious National Academy of Sciences and a recipient of its highest honor. Interestingly, Duesberg was hailed as one of the brightest lights in the biomedical firmament by those who are now his critics in the "AIDS" establishment. I have also been influenced by the work of Dr. Luc Montagnier, the discoverer of HIV, who no longer believes that HIV is the sole cause of "AIDS."

These two dissenting voices are not alone. Hundreds of scientists and specialists have organized a group for the "Scientific Reappraisal of the HIV/AIDS Hypothesis," and they are locked in battle against the NIH's research lobby. Their purpose is to refute the self-serving pronouncements of the "AIDS" industry.

Dr. Kary Mullis, winner of the Nobel Prize for inventing the first test to detect HIV and other latent viruses, is among those who have concluded that HIV does not cause "AIDS." "I went around asking a lot of people what the references were that HIV caused AIDS, but no one could tell me,"[20] he said in one report. Finally Mullis tracked down Dr. Montagnier, who could only tell him to "quote the monkey work."[21]

"And then it dawned on me," Mullis said. " 'Oh, God, there isn't such a paper!' "[22]

At a conference on "AIDS" in 1992, thirty cases of non-HIV "AIDS" were reported. These were people with full-blown "AIDS" with no HIV in their bodies. Call it HIV-negative "AIDS." This phenomenon stands the HIV-causes-"AIDS" theory on its head and prompted the *New York Post* to acknowledge: "The incidence of such cases supports a small but growing body of expert opinion which holds that HIV does not itself actually destroy the body's immune system. Adherents to this view believe that AIDS results from a combination of factors—including, in most cases, HIV."[23]

In general, the American media have fallen for the scam of the virus hunters and merely parroted the undocumented theory of "AIDS" causation. Major elements of the British press, on the other hand, have challenged the "AIDS" industry. In 1987, the *Times* of London reported that it was the World Health Organization's smallpox vaccine that "triggered the AIDS virus" in Africa.[24] This fascinating story by one of the most prominent news sources in the world was completely ignored by the American media.

Some of the epidemic-and-plague hysteria on "AIDS" is the result of stories by ill-informed reporters manipulated by the "medical CIA," the Epidemic Intelligence Services (EIS) of the CDC. Some one hundred elite EIS agents have been graduated each year since 1951 from the CDC's intelligence service, and many have become prominent medical writers. Others go to World Health Organization (WHO) agencies in the United States and other countries or to biotechnology companies, foundations that fund medical research, and scholarly journals. Each of these institutions is a player in the "AIDS" establishment, and each is overtly hostile to any opposition to the HIV-causes-"AIDS" hypothesis. It is noteworthy that every member of the CDC-EIS network remains a part of the Public Health Service and subject to recall to active duty in the event of war.[25]

The Cure That Kills

The $100-billion-a-year War on "AIDS" has not cured one person or saved one single life, but the most cited treatment for "AIDS," the drug AZT, is increasingly recognized as a killer itself.[26] If you get HIV, you get AZT and "AIDS." And 75 percent of prescribed AZT is used by Americans. Dr. Duesberg has characterized the prescription of AZT for "AIDS" patients as "AIDS by prescription."

"It is a terminator of DNA, which is the central molecule of life," Duesberg said on my PBS television show. "If you terminate DNA synthesis, you terminate life. There is no way around it. It is as straightforward as cyanide."[27]

No AZT for Gallo

The government's star "AIDS" researcher, the chief promoter of the HIV-causes-"AIDS" theory, and the champion of AZT therapy for HIV infections, Dr. Robert Gallo, reportedly said that if he were infected with HIV, it would take twenty to thirty years for him to develop "AIDS." When and if he did develop it, he suggested, he might not take AZT, according to Dr. Peter Dusenberg and Dr. John Yiamouyiannis in their book *AIDS*.[28]

AZT was designed almost thirty years ago to treat leukemia by "termination of DNA synthesis,"[29] Duesberg said. When AZT failed to prolong the lives of leukemia animals and was not accepted for cancer chemotherapy, it was shelved until 1987, when understandably panicked homosexuals cried for the quick development of an "AIDS" drug. But treating the HIV virus with AZT amounts to incredible and dangerous overkill, Duesberg said. "With AZT you have to kill 499 uninfected cells every time you want to kill an infected cell. That's what's called an extremely high toxicity index. That is, if you want to get one criminal, you shoot five hundred people,"[30] Duesberg explains.

The list of AZT critics grows each year. The four-year Concord double-blind placebo controlled study of AZT, published in *Lancet* in April 1993, "suggested AZT offers 'no significant benefit' to infected, healthy people in slowing the disease ["AIDS"] or prolonging life."[31] In the widely respected study, it was found that AZT produced no benefit in delaying disease progression or in preventing death. Before he died of Kaposi's sarcoma, "AIDS" activist Michael Callen said, "I firmly believe that AZT not only does not extend life, but, in many cases, shortens life."[32] Callen had this message to the homosexual community: "We blew it. We've murdered our own people."[33]

Michael Callen was vilified by some who disagreed with him, but he was venerated by others for his courage in challenging the medical establishment. He was honest. Unfortunately, others are not. Despite increasing scientific evidence that AZT does not help and in fact may kill people, the medical establishment is still adamant about administering AZT as a treatment for "AIDS," particularly for the children born to "AIDS"-infected mothers.

AZT certainly did not save Steve and Cheryl Nagle's daughter Lindsay, as ABC's Forrest Sawyer noted when his *Day One* television show profiled the Nagle family in a segment entitled "AZT."

"The government and the company that developed AZT campaigned to get patients on the drug as soon as they were diagnosed, but it turns out that may have been a hasty and perhaps even deadly course of action,"[34] Sawyer revealed.

Reporter John Hockenberry of *Day One* explained that the Nagles believed what their doctors told them—that "they had something that fights HIV," namely AZT. The family accepted the treatment because they believed their doctors also when they told the family that "if you test positive for HIV, it's only a matter of time—AIDS and then death."[35]

For two years, beginning at three months of age, the "completely healthy" baby was given AZT, and the results were "dramatic." The parents would be awakened during the night to the girl's "shrieking in pain." Leg cramps started, she lost muscle tone, and her legs and arms "just turned into flab,"[36] the father told *Day One*.

The Nagles came to believe, as many others have, that the treatment for HIV was far worse than the disease it was alleged to cause. The Nagles took little Lindsay off the AZT and her leg cramps stopped, her appetite returned, and she gained weight, and, Steve Nagle said, "She stopped screaming in the middle of the night."[37]

In June of 1995, almost two years later, Hockenberry told me that Lindsay had actually converted to HIV-negative after getting off AZT. Later in this chapter you will read of a young lady, Kimberly Bergalis, who stayed on this deadly drug and died.

Another therapy is being promoted by the Nation of Islam (NOI) as a "miracle drug" that cures "AIDS." In reality, it is nothing more than an illegally dispensed interferon called Kemron that is potentially dangerous and extremely costly. The NOI charges a $1,000 initiation fee and sells Kemron for $150 to $250 for a monthly supply. The same supply can be purchased for only $55 at other outlets.

The True Causes of "AIDS"

The truth is, we do not have a cure for "AIDS" because we really don't know what causes it. The U.S. government's leading "AIDS" researcher, Dr. Robert Gallo, claimed in 1984 that HIV is the only cause of "AIDS." But Dr. Duesberg, whom Gallo once boasted "knows more about retroviruses than any man alive,"[38] discounts Gallo's theory that HIV alone has the ability to do that. Laurie Garrett's book discredits Duesberg, makes the erroneous allegation that "evidence for the cause of AIDS" exists, and inaccurately characterizes Duesberg's argument.[39] Indeed, the *New York Newsday* science reporter admitted to a bias in the August/September 1994 issue of *Poz* magazine: "It would be naive of me to think that all my reporting on AIDS is objective. . . . When I write, 'HIV, the virus which causes AIDS,' that is not an absolutely neutral choice."[40] This kind of media conformity to government propaganda notwithstanding, Duesberg and many others persist in their belief that drugs—and the treatment AZT itself—are at the root of this controversy.

"AIDS" in the United States and Europe "is the consequence of the ever-increasing consumption of psychoactive drugs, and last, but not least, AZT, which is used to inhibit HIV, but is itself the most direct cause of AIDS,"[41] said Duesberg in an appearance of *Tony Brown's Journal*.

In Short, Duesberg's theory of "AIDS" is that it is a collection of twenty-nine known diseases that occur in the presence of HIV antibodies. Without HIV, these diseases go by their old names. With HIV present, doctors call these old diseases "AIDS." Duesberg and his supporters believe now that whatever you call them, the "AIDS" diseases are likely caused by the use of anti-HIV and recreational drugs rather than solely by the human immunodeficiency virus. In hemophiliacs who contract "AIDS," he blames impurities in transfusion fluids. Duesberg blames the rise of "AIDS" on the "massive escalation in the consumption of recreational drugs" in the 1960s and 1970s. In a ten-year period alone, Americans increased their use of cocaine by 200 percent, while the use of amphetamines and poppers skyrocketed among homosexuals.[42] Drug abuse, Duesberg says, resulted in the reemergence of old diseases such as tuberculosis—one of the "AIDS" diseases—in the 1980s and 1990s.[43]

If that is the case—and it is a much more valid theory than the one claiming that HIV alone causes "AIDS"—I believe we have given the wrong name to this syndrome. We should call it "Drug Acquired Immunodeficiency Syndrome," or "DAIDS."

Duesberg's theory of how "AIDS" spreads is simple to follow. He holds that "AIDS" begins in those who are biologically most susceptible: people whose lifestyles make them perfect hosts for a "benign" retrovirus (HIV). He says that HIV hardly ever becomes active even in "AIDS." These "thirdworldized" hosts, all of whom have ravaged their body in some way, include heterosexual drug addicts; homosexual and bisexual men who are also drug addicts; and those drug-abusing homosexuals whose irresponsible "bathhouse" sex behavior exposes them to lethal microbes and the spread of infections.

Michael Callen, one of the founders of the People with AIDS Coalition, lived twelve years with "AIDS." Just before his death, he offered a compelling confession in *HEAL*, a publication for alternative health therapies, that lends credence to Duesberg's "DAIDS" theory:

By the age of 27, I estimate I had had 3,000 different sex partners. I'd also had: hepatitis A, hepatitis B, hepatitis non-A, non-B; herpes simplex types 1 and 2; shigella; entamoeba histolytica; Giardia; syphilis; gonorrhea; nonspecific urethritis; chlamydia; venereal warts; CMV; EBV reactivations; and finally cryptosporidiosis and AIDS. The question for me wasn't why I was sick with AIDS but rather how I had been able to remain standing on two feet for so long. If you blanked out my name and handed my medical chart, prior to AIDS, to a doctor, she/he might reasonably have guessed that it was the chart of a 65-year-old equatorial African living in squalor.[44]

Callen very likely put his finger on what is the probable link between the "AIDS" outbreak among high-risk groups in the West and the malnourished heterosexual population in Africa: a Third World health status. This sort of ravaged immune system first developed in the West among a bathhouse culture of male bisexuals and homosexuals, as well as heterosexual injection-drug users and homosexual long-term recreational drug abusers. These subcultures were extremely vulnerable because of debilitated bodies and a Third World hygiene status.

Callen's truthful portrayal of a dangerous lifestyle that some, but certainly not all, homosexuals practice demonstrates that "AIDS" has attacked them because they are susceptible—not because of their sexual orientation. Many scientists have come to believe, as Callen did, that the HIV virus is normally benign and usually causes no problem in reasonably healthy individuals.

"Something other than homosexuality" causes "AIDS," says Duesberg. "Your all-American homosexual neighbor will never get 'AIDS.' It's only the ones who have hundreds, or thousands, of sexual contacts a year. And how is that achieved? Almost exclusively by chemicals.[45]

"Drug abuse is rampant among homosexuals who practice promiscuous sex. For multiple orgasms and as an anal relaxant, this bathhouse subculture routinely uses 'poppers' (amyl nitrite inhalants), and their 'recreational' regime consists of PCP, amphetamines, angel dust, cocaine, heroin, uppers and downers, Valium, and alcohol,"[46] Duesberg explains.

Among the unhealthy fallout from poppers is Kaposi's sarcoma, which is now classified as one of the "AIDS" diseases. "Kaposi's sarcoma hits the face and the lungs because it's inhaled—the pulmonary Kaposi's sarcoma,"[47] Duesberg says. Kaposi's sarcoma shows up almost exclusively among homosexual men, not hemophiliacs or junkies. Duesberg says there are only two ways a selective epidemic (admittedly an oxymoron) such as "AIDS" could appear in the population: through either an infectious agent or a toxic substance.

"If it's an infectious agent, everybody can get it. For example, you can only get a tennis elbow if you are an active tennis player. And you almost only get liver cirrhosis if you are a heavy drinker. So you can distinguish an infectious disease from a noninfectious disease by looking at who gets sick. That's called epidemiology."[48]

"AIDS" is not infectious. It is "acquired." Therefore, certain groups must be doing something the rest of the population is not doing. One of those high-risk behaviors is drug usage. In his book *AIDS: The HIV Myth*, Jad Adams quotes

Larry Kramer's 1978 novel *Faggots*, which lists a plethora of drugs commonly used by some homosexual men. In addition to nitrous oxide, the active component of the ubiquitous poppers, the startling array includes MDA, MDM, THC, PCP, STP, DMT, LDK, WDW, LSD, cocaine, mescaline, angel dust, Benzedrine, Dexedrine, Dexamyl, Desoxyn, strychnine, Ionamin, Ritalin, Desbutal, Opitol, ethyl chloride, nitrous oxide, crystal methedrine, Clogidal, Nesperan, Tytch, Nestex, Certyn, Preludin with B-12, Zayl, Quaalude, Tuinal, Nembutal, Seconal, Amytal, phenobarbital, Elavil, Valium, Librium, Darvon, Mandrax, opium, Stidyl, Halidax, Calcifyn, Optimil, and Drayl.[49]

When "AIDS" first publicly appeared, the early "AIDS" cases were found spreading among a homosexual subculture that was hepatitis-riddled, drug-poisoned, sexually-transmitted-disease-laden, and immune-depressed. These were participants in a bathhouse sex scene that today most homosexuals view as irresponsible and dangerous.

It has been strongly argued that "AIDS" industry scientists knew that the first waves of "AIDS" victims were mostly drug addicts and homosexual drug abusers, but they covered up the toxicological correlation to drugs because it did not fit their potentially enriching theory that it was a viral-based disease. From this first small group of homosexual men, the old opportunistic diseases (which came to be known as "AIDS" in their revived forms) enjoyed a resurgence and spread in an ever-widening circle via the hosts' sexually intimate partners, who may also have shared drugs. Again, those at highest risk were those in poorest health with the most dangerous behaviors—bathhouse men and intravenous-drug users (IVDUs)—a process now known as "thirdworldization." These two subsets became the primary "AIDS" groups— and still are after fifteen years of a "confined epidemic" (another oxymoron).

As proof of this theory, consider the probability that drug addicts were dying of "AIDS" diseases long before any- one looked for HIV or even for a cause of death among injec- tion-drug abusers and before the outbreak among the homosexual subset of recreational-drug addicts.

To buttress this correlation between drugs and "AIDS," author Laurie Garrett cites a study by the U.S. National In- stitute of Drug Abuse in which intravenous drug users tested HIV-positive as early as 1971–1972. "It was the druggies not the gays who started it," Garrett quotes virologist William Haseltine as saying.[50]

By 1981, when homosexual men were first identified with "AIDS," 30 percent of intravenous drug users were al- ready HIV-antibody-positive. And Garrett notes that in the 1970s, homosexual men were "two or three times more likely" to be intravenous drug users.[51] A large proportion of the initial cases of men with "AIDS" were identified as both drug abusers and homosexuals, she writes.

Hemophiliacs got the virus through transfusions of blood contaminated not only with HIV, but with hepatitis, syphilis, herpes, and so on. Babies born to IV-drug-abusing prostitute mothers also became nonacquiring or secondary victims.

In my opinion, then, the illness we call "AIDS" in the United States is not by any means a "homosexual disease." I believe it is precipitated by a chemical injury, but it is also triggered by a variety of microorganisms as cofactors that destroys the body's immune system. The process is a deadly synergistic combustion. High-risk "AIDS" behavior in the West is primarily drug abuse, receptive anal intercourse, poor hygiene, malnutrition, and unprotected sex—especially if there is a history of sexually transmitted diseases.

The Burden of Disease

This theory makes sense when you consider that in Third World countries, "AIDS" first appeared not among the drug addicts and homosexuals as much as it did among those who experienced poor sanitation, malnutrition, poor medical care, and years of being the unwitting victims of experimental and often contaminated vaccines—often those vaccines that are dumped in Africa and other poor nations by drug companies when the vaccines get near their expiration date.

Speaking of the Western hostility toward Africans, in a February 2, 1987, article entitled "AIDS in Africa," the *Guardian* described prostitutes near Nairobi, Kenya, each of whom serviced an average of one thousand truck drivers each year.[52] In 1980, the *Guardian* explained, none of the six hundred African women tested HIV-antibody-positive. By 1983, a significant 53 percent were positive. The percentage increased again in 1987, to 80 percent who were infected with the "AIDS virus," the *Guardian* reported.

However, Richard and Rosalind Chirimuuta, authors of *Aids, Africa and Racism*, point out the obvious. "AIDS" was announced in the United States in 1981, and the "AIDS virus" (HIV) and the blood test to find it came in 1984. Therefore, how could Third World, medically backward Kenya test for a virus the world had never heard of in 1980, using a testing instrument that would not be introduced to the world until 1984 at the same time the virus appeared?[53]

African men and women are predisposed to "AIDS" by environmental factors, while drug use is the primary predisposer in the United States. I suggest that starvation and malnutrition in Africa are what we call "AIDS." And that chemical poisoning from long-term psychoactive-drug consumption in the United States is "DAIDS." In distinguishing "AIDS" in Africa and "DAIDS" in the western hemisphere from disease in general, there is an essential difference. Hot climates are not good for your health, and "the worst of these climates, in terms of human disease, is tropical Africa,"[54] according to *Camping with the Prince and Other Tales of Science in Africa*, by Thomas A. Bass. The African continent is a "burden of disease,"[55] says Bass. "Nowhere in the world is the mass of parasites, fungi, bacteria, and viruses heavier than in Africa because its landmass straddles the equator. This makes Africa hot and wet, a perfect climate for disease [that provides a] year-round community of arthropods and other disease-spreading organisms."[56]

Bass's book contains a description of the work of Dr. Oyewale Tomori, a Nigerian virologist, who explains that the principal cause of immune deficiency and death in the "AIDS belt" of Africa is not a virus, but malnutrition. I strongly recommend a careful reading of the following paragraph by Bass, a White American, to purge the endemic American and European hatred and misunderstanding of this continent:

> It is generally not the exotic parasites associated with tropical medicine that kill people in Africa, but the common bacterial and diarrheal diseases and respiratory infection that cause two-thirds of the

deaths in the Third World. Retroviruses are one cause of diarrhea, while influenza and adenoviruses are implicated in respiratory infections. Malaria comes next on the list, as the primary or contributing cause for a third of the deaths in sub-Saharan Africa, which is the epicenter of [AIDS]. But as public health officials know, the underlying cause of death in most of these cases is not a virus or a parasite. It is an acquired immune deficiency syndrome known as malnutrition.[57]

Bass's theory explains the heterosexual "AIDS" epidemic in Africa, and it is consistent with the *Merck's Manual* explanation in 1952 of the causes of "acquired immune deficiencies" in the United States as malnutrition and drugs.[58]

The Lies About Heterosexuals and "AIDS"

The myth that heterosexuals are at risk from "AIDS" has been spread largely by homosexual leaders, who mistakenly believed that if the public was panicked into fearing that everyone was in danger, more funds would pour into "AIDS" research. They adroitly outmaneuvered the campaign to label "AIDS" as a "homosexual disease" by selling another lie. They promoted "AIDS" as a heterosexual disease as well.

The truth is, "AIDS" is neither; it is mostly a drug-precipitated disease. The chance of "contracting AIDS from a heterosexual without risk behaviors" is about one in one million—the probability of winning a state lottery or being struck by lightning, according to Dr. Robert Root-Bernstein, an associate professor of physiology at Michigan State University and author of *Rethinking AIDS*. This means that usually in primary cases of "AIDS," one must do something to "acquire" it, and "HIV is extremely difficult to transmit to a healthy individual,"[59] the physiologist says.

Many people simply do not know what "risk behaviors" mean. Inane and trendy phrases such as "safe sex" and "mixing bodily fluids" have caught on, but sex education in the age of "AIDS" unfortunately remains an oxymoron.

Another point of confusion has to do with the effectiveness of condoms. Rush Limbaugh was chastised by the editor of *EXTRA!*, a publication of FAIR (Fairness and Accuracy in Reporting), for saying, "The worst of all of this is the lie that condoms really protect against AIDS. The condom failure rate can be as high as 20 percent."[60]

FAIR called him "plain wrong" and retorted with its own facts: "A one in five AIDS risk for condom users? Not true, according to Dr. Joseph Kelaghan, who evaluates contraceptives for the National Institutes of Health. 'There is substantive evidence that condoms prevent transmission if used consistently and properly,' he said. He pointed to a nearly two-year study of couples in which one partner was HIV-positive. Among the 123 couples who used condoms regularly, there wasn't a single new infection."[61]

I interpret that to mean that he is saying that condoms prevent HIV infection 100 percent of the time.

There were actually two "blockbuster" studies on condoms—both focusing on heterosexual couples who had sex for two years, with the male partners carrying HIV. In the Italian study, among consistent condom users, only 2 percent of the women were infected. In the European study group, where the average couple had sex 120 times, among inconsistent condom users, 10 percent of the women became infected. When condoms were always used, HIV was never passed to the women.

However, a 1992 study published by the FDA's Division of Physical Sciences, Center of Devices and Radiological Health, which was commissioned and financed by the CDC, seems to support Limbaugh. "Of the condoms tested, 29 (32.5%) leaked the fluorescent dye. . . . Hence, our results support the conclusion that the use of latex condoms can substantially reduce, but not eliminate, the risk of HIV transmission,"[62] said the authors of the FDA study.

Another aspect of risk comes to mind. Suppose one person's HIV status is unknown, but he or she does not belong to any high-risk group (homosexual or bisexual male, hemophiliac, blood-transfusion patient, or drug abuser). The risk then sinks to only one in five million—that's one in fifty million per act of intercourse. However, studies suggest that women who engage in anal intercourse (as well as the receptive partner in homosexual anal sex) are at greater risk for HIV infection. Dr. Root-Bernstein says that anal intercourse is dangerous for the passive partner, but almost risk-free for the anal penetrative partner, and oral-anal intercourse (anilingus) carries risk for the oral partner regardless of sexual orientation.[63]

Not surprisingly, women who are active in anal intercourse also have a higher incidence of additional infections, such as cytomegalovirus, chlamydia, shigella, and amebiases. Root-Bernstein interprets these studies to mean that "heterosexual anal intercourse and a history of sexually transmitted diseases but not promiscuity per se confer the same increased risk of AIDS as has been found among homosexuals."[64]

The occurrence of HIV and "AIDS" among non-drug-abusing women is very small, especially when you do not include the new indicator diseases that were added by subsequent "AIDS" redefinitions to increase the number of female heterosexual cases for political purposes. Overall, only 6 percent (10,011 cases) in the United States in 1991 were heterosexual "AIDS"—and the majority of cases involved sex with someone from a high-risk group: IV-drug user, bisexual male, hemophiliac, transfusion patient, or, Root-Bernstein explains, a "person from a country in which AIDS is widespread among heterosexuals."[65]

It is sex with IV-drug abusers that results in over two-thirds of "AIDS" cases among heterosexuals.

Transmission of HIV from women to heterosexual men is rare and difficult; it is the most unlikely route of infection, occurring in only 2.2 percent of the cases. Even female prostitutes "represent virtually no risk for spreading HIV to nonrisk heterosexuals."[66] Non-IV-drug-abusing female prostitutes "almost never become infected with HIV," and "even drug-abusing, HIV-positive prostitutes" rarely infect non-drug-using heterosexual men.[67] Neither frequency of sexual intercourse nor oral intercourse is high-risk behavior under normal circumstances. Only in rare cases has receptive oral intercourse (cunnilingus and fellatio) with ejaculation resulted in HIV infection.[68] Saliva, a study reports, is a brake against the transmission of HIV.

"AIDS" and Magic

Even though heterosexuals are rarely at risk, there has been a nationwide campaign to spread fear of "AIDS" among heterosexuals. And at the forefront are stories headlined "If

Magic Johnson Can Get AIDS, Anyone Can Get AIDS."
While Michael Callen became a symbol that people can live
with HIV because of the truth, the NBA star became a poster
boy for dying people and a symbol of celebrity "AIDS" based
on a scientific lie.

If he really wanted to help, Johnson, who is HIV-anti-
body-positive but does not have "AIDS" as the headline
falsely claimed, could have emphasized his good health and
ability to continue to play basketball, thus demonstrating
that HIV antibodies are not a death sentence. It appears that
Johnson has finally come to this realization. Dr. Duesberg
has stated that in most cases, HIV is not harmful because
the body has vaccinated itself against HIV. "When we were
young, we were artificially inoculated with attenuated vi-
ruses—polio virus, measles virus, mumps virus—to elicit
that response. Once you have made those antibodies, you
are protected for the rest of your life. Once you have antibod-
ies, you are protected, you are vaccinated."[69] Duesberg is as-
tonished that medicine has suddenly identified a natural
vaccination as a death sentence. The medical field even calls
this natural defense mechanism an "HIV disease."

With all respect to Duesberg's brilliance, there is the
possibility that in the case of HIV the body's antibodies do
not destroy the virus because they may be nonneutralizing
antibodies. If this is true, the body has neither neutralized
nor destroyed the HIV.[70] In my opinion, this is likely to be
the case because HIV is present at so many disasters—at
least as a co-factor. Whether or not one agrees that HIV alone
causes "AIDS," HIV poses an immediate or potential danger,
in my opinion.

These facts cast doubt on the heterosexual-scare cam-
paigns of the CDC's most prominent symbol for the so-

called "AIDS" epidemic, Johnson. There is also the political reality that some homosexual leaders and a sympathetic media were using the basketball star's predicament to create a famous heterosexual symbol for "AIDS" in order to attract more empathy and money for "AIDS" research. If this is true, in many ways this strategy backfired.

The homosexual activists may have inadvertently hurt their own people with this hypocrisy, because the heightened concerns—and potentially much larger range of victims—resulted in more funds pouring into the "AIDS" establishment's corrupt research programs. It also boosted sales for AZT, an extremely profitable anti-HIV drug that, as I have noted, does not cure "AIDS" and may only worsen the health of those who take it.

The Bergalis Case

The most persuasive case made by the CDC disinformation machine in its effort to prove that the average heterosexual is at risk for "AIDS" was that of Kimberly Bergalis, a twenty-two-year-old White woman who was advertised as the virgin victim of an evil homosexual dentist who, it was suggested, may have intentionally transmitted the "AIDS virus" to her while extracting two molars.

Her dentist had Kaposi's sarcoma, typical of homosexuals with "AIDS." Bergalis had a yeast infection. Both diseases were called "AIDS" because of the presence of HIV. Bergalis said she had never used drugs and had never had sex. She testified before Congress, "I did nothing wrong, yet I'm being made to suffer like this."[71] However, after she was subpoenaed for a gynecological examination, human papilloma virus (HPV) type 18, "a strain known to be sexually transmitted and highly associated with HIV in gay and bisexual men," was found. This and the perianal warts that were also reportedly found are almost irrefutable evidence of sexual intercourse.[72]

Before she died, Bergalis finally admitted to having been sexually active, and it is not unreasonable to suspect that anal intercourse with a bisexual man was a part of her high-risk behavior.

About 10 percent of heterosexual women and, studies suggest, 25 percent of teenage girls engage "in rectal intercourse . . . to avoid pregnancy or to retain their virginity."[73] But the risk for HIV infection and "AIDS" is the same as for females who do not practice anal intercourse—almost nonexistent.

We are still left without a valid explanation for the rapid deterioration of Bergalis. She suffered severe anemia, weight loss, balding, and muscular wasting. All of these symptoms occurred within a few weeks of her taking AZT. Soon after the treatment, she was restricted to a wheelchair, and then she died. Can a yeast infection do all of that? Or was Kimberly Bergalis sentenced to death because she had an HIV infection?

Lies About Blacks and "AIDS"

In 1986, the CDC suspiciously added Blacks born in the "Pattern II" countries (Haiti and Central Africa) when counting heterosexually transmitted "AIDS" cases among Blacks in the United States. This accounting technique had a great impact. It doubled the total number of all heterosexual

"AIDS" cases in the United States because "Pattern II" countries have a majority of heterosexual cases. When added to the U.S. Blacks-only category, it fueled the illusion of a dramatic explosion of "AIDS" among U.S. Black heterosexuals. According to one media report, half of the "AIDS" cases among heterosexual Blacks in Detroit are among the Black sub-group from Pattern II countries—mostly Haiti and Central Africa.[74]

This government maneuver expanded heterosexual cases among Blacks significantly and skewed the epidemiology to suggest that the ordinary Black person is increasingly "acquiring" these so-called "AIDS" diseases through an escalation in high-risk behaviors.

If you delete the additional "Pattern II" African groups and count only Black Americans, you can see that the major HIV and "AIDS" health threat to Blacks is also coming from injection-drug users. Once the "Pattern II" groups are deleted, heterosexual "AIDS" cases among Blacks, two-thirds of which result from sexual contact with IV-drug users, jump from 25 percent to 60 percent because the deletion exposes the artificial inflation of non-drug-using heterosexuals in the accounting. Therefore, it becomes more representative of the behaviors of American Blacks for good or bad. This offers a more accurate picture of a drug-abusing "AIDS" subset group among Black Americans that disproportionately accounts for "AIDS" in the Black community.

The numbers hardly describe Black America as the number one "AIDS" community or Blacks per se as a high-risk group. Of 32 million Black people, including the Pattern II groups, as of December 1993, only 44,936 or 0.14 percent (one seventh of 1 percent), had "AIDS." Of that number, 14,828, or 33 percent of the Black "AIDS" cases, are homosexual or bisexual men. And while only 7 percent of Black "AIDS" cases are homosexual IV-drug abusers (7 percent for Whites in this category), 38 percent are heterosexual IV-drug abusers (7 percent for Whites).[75] Black drug abusers—homosexual (7 percent) and heterosexual (38 percent)—total 42 percent of the total Black population with "AIDS." This compares to 7 percent among Whites for both homosexuals and heterosexuals in these categories for a total of 14 percent. The use of drugs and drug injury are consistent indicators of why Blacks, 12 percent of the U.S. population, constitute 36 percent of all "AIDS" cases. Drug users among Blacks and Hispanics, a combined total of 21 percent of the U.S. population, account for 54 percent of this country's "AIDS" cases—a figure strongly impacted by injection-drug use.

And finally, after years of Black-equals-"AIDS" propaganda, the National Center for Health Statistics released a report in early 1995 noting that only about 2 percent of deaths among Blacks in 1994 were from "AIDS." It listed the leading causes of death among Black Americans to be the same as among White Americans—heart disease, cancer, and strokes. These three ailments account for over 60 percent of all deaths, according to the report. "AIDS" is the leading killer of Black males between 25 and 44 years old, but most of those deaths are among male homosexuals and IV drug users. Outside of those two populations, "AIDS" has yet to make any significant headway into the rest of the Black population, the study said.[76]

"AIDS" and Scientific Racialism

As I mentioned at the beginning of this chapter, Blacks have a particular concern about the lies that surround "AIDS," specifically those lies that somehow try to pin the disease on Blacks and those that suggest that "AIDS" is somehow partial to Blacks.

In his book *The Myth of Heterosexual AIDS*, reporter Michael Fumento played the race card in alleging that Magic Johnson was at greater risk for an HIV infection than a "heterosexual white basketball player" because of Johnson's "opportunities to have intercourse with inner-city black women."[77] In 1992, Fumento revisited this racialist sociology from his book in an article in the *American Spectator*: "AIDS the disease makes distinctions based not only on sexual habits but also on race."[78]

Fumento offered not one shred of scientific evidence linking the genetic traits of Americanized Africans or Hispanics to a susceptibility to HIV or "AIDS." In fact, there is none. Fumento's hypothesis that nature singles Blacks out for this disease is borrowed from the perniciously racist epidemiology of the U.S. government's disinformation machine, the CDC, and the "AIDS" establishment's National Commission on AIDS.

In some instances, there are biological risk factors that cannot be changed. Studies have found, for example, that Blacks have a much greater risk of death and disability from stroke than Whites—about 60 percent higher—because Blacks have a higher incidence of high blood pressure. But to say, as Fumento does, that the thirty "AIDS" diseases make racial "distinctions" smacks of prejudice.

Fumento's racial distinctions theory is shared by

psychologist-turned-evolutionist J. Philippe Ruston, who in his book *Race, Evolution and Behavior* said that Blacks have a disproportionate number of "AIDS" cases because they are genetically programmed to be sexually promiscuous and thereby more prone to get "AIDS."[79]

If Fumento knew that "AIDS" is the leading cause of death among young men, especially among those twenty-five to forty-four years old, in Italy, I wonder if he would rush to the judgment of racial "distinction." Or perhaps he would accept the more rational explanation that the phenomenon may result from a very high incidence of psychoactive drug use among young Italian men. As Dr. Duesberg points out, 72 percent of "AIDS" occurs among American males aged twenty to forty-four because they constitute over 80 percent of psychoactive drug users.[80]

Fumento poured salt on this open wound of racism when he boasted that "to the extent that AIDS is a heterosexual disease it is one not so much of whites . . . but rather of inner-city blacks and to a lesser extent Hispanics, especially Puerto Ricans."[81]

"Black men," Fumento inaccurately claimed in a 1992 article, "have 50 times the chance of getting AIDS as whites."[82] He bases that astonishing conclusion on the basis of 3,146 heterosexual Black men who have "AIDS." That's 3,146 out of about sixteen million Black men, and they are mostly needle-sharing slum dwellers with a history of sexually transmitted diseases and severe multiple infections. The truth is that Black men, on average, are *five* times more likely to get "AIDS" than Whites, according to a 1993 CDC report, not Fumento's fifty times.[83]

In an interview on my television show, Fumento backpedaled on his racist claims. "You're right, Mr. Brown," he admitted, "it's not a matter of skin color. There's nothing genetic to predispose you to AIDS."[84] But he contradicted this admission two years later in his book: "Even a middle-class black is at increased risk of sexually transmitted AIDS, for the simple reason that middle-class blacks have a better chance of receiving the virus from a lower-class black than do whites for a lower-class white."[85]

Fumento is both disingenuous and culturally wrong. Drugs are culturally more acceptable among homosexuals as a lifestyle choice, and many homosexuals are advocates of anal sex as a natural form of sexual intercourse. This cultural approval is not present among mainstream Blacks and Hispanics, who consider drug users pariahs and do not condone anal intercourse.

This racist-inspired epidemiology judges Blacks and Whites by a double standard that casts Blacks in an unfavorable light as the reservoir of "AIDS" in America. This stereotype will drive the adoption of public policies singling Blacks out as the cause of the so-called "AIDS" epidemic and isolate them, along with Hispanics, as America's "AIDS" community. If that happens, these marginalized groups become easy targets for triage if and when the majority of the population comes to feel threatened by the spread of "AIDS."

If the "AIDS" establishment's disinformation succeeds in identifying a scapegoat as America's "AIDS" community, it can then quarantine all of its members, take away their jobs or refuse to hire them, or, worst of all, legally impose upon them extremely toxic antiviral drugs, as is now being done with HIV-antibody-positive pregnant women, their fetuses, and their newborn babies. The majority of these women, fetuses, and babies are Black, and many of the women are IVDUs, often prostitutes. That's how triage works.

Fumento claims that Blacks like myself—and he refers to me by name in his book—suffer from upper-class oversensitivity, proving that his estimation of Black intellectual potential could never be understated. He is only a little less dangerous than the middle-class Blacks who are regurgitating the CDC-led campaign that explains a pernicious racism as "AIDS" epidemiology and "AIDS" prevention. While Blacks are dying from the government's recommended "AIDS" toxic treatment, Black elitists are using the increased death rates as proof that Blacks need more AZT treatment. Their slogan should be "Save me a seat on the next boxcar to Auschwitz!" Drugs, not exotic diseases, are killing Black people. If it were not for the devastating presence of drugs, the CDC-led "AIDS" disinformation campaign against Blacks would crumble.

In 1994, Dr. David Satcher, a Black man and a friend, was appointed head of CDC. He is a man of impeccable integrity and honor, as well as a brilliant scientist and administrator. He has assisted scores of young Black men and women in getting medical educations. He has no connection to any of CDC's rogue behavior. I wish my friend well and I trust him, but I am distrustful of the "AIDS" industry and all of its public and private arms, especially the private "AIDS" commissions it manipulates. Dr. Satcher would be well-advised to watch his back.

Triage #ACTG 076

When I interviewed the head of pediatrics at a major hospital, he advocated AZT for pregnant HIV-antibody-positive women, fetuses, and newborn babies to prevent transmission of the HIV to the babies. This is also the recommendation of the questionable clinical trial ACTG 076 by the National Institutes of Health. This trial promises that HIV-infected women can reduce the transmission of HIV antibodies to their newborns by ingesting AZT.

This government-imposed directive seems even more backward when you examine the case of the HIV-infected five-year-old who astonished doctors in Los Angeles by naturally clearing his body of HIV. According to *The New England Journal of Medicine*, the boy tested positive twice before testing negative at the age of twelve months. As a matter of fact, three out of four babies born to HIV-positive mothers are born without HIV.

The HIV-causes-"AIDS" theory kicks in and assumes that within ten years the mothers and the children will develop "AIDS" and die. Therefore, the government insists, it is safer to feed the fetus AZT, a very toxic drug that kills HIV-infected cells along with healthy cells, during the final trimester of pregnancy. Only 19 percent of the women in this study were White; 51 percent were Black and 29 percent Hispanic. That is an 81 percent non-White sample, which is close to the ethnic distribution for pregnant women in this drug-related group.[86]

"Remember, AZT was made available thirty years ago for one purpose only: to kill human cells, to terminate DNA synthesis in growing human cells, including cancer cells. To put that into babies? That is beyond me,"[87] said a shaken Dr. Peter Duesberg when discussing this trial. It clearly constitutes an out-of-control research-medical-drug establishment scrambling for fame and profit on the unproven theory that HIV causes "AIDS."

Of the six thousand to seven thousand women with the HIV antibodies who give birth, about fifteen hundred to two thousand of their children are born with the HIV antibodies.[88] That means that if left alone, without toxic chemicals, about 70 percent will *not* be born with the HIV antibodies anyway. It is "irresponsible," Duesberg says. "Even if you had a paper that showed that HIV causes AIDS, where are the data that show that AZT is not hurting the rest of the baby?"[89]

The federal Public Health Service and its "Tuskegee experiment" cohort, the CDC, Senator Jesse Helms, and a coalition that includes religious groups are vigorously promoting prenatal HIV counseling for the purpose of convincing pregnant HIV antibody positive women (including those who are healthy but will test false positive) to transfer AZT to their fetuses. A positive test is not evidence that the baby is infected—just that antibodies from the mother are in the infant's blood. Over a period of time, 70 percent eliminate them in their urine.

But those facts didn't stop CDC Director David Satcher from citing studies that he said proved AZT saved babies' lives and was "an unprecedented breakthrough in HIV prevention."[90]

The *Times* said that, in a study of 839 children, the drug AZT "proved so ineffective in preventing disease [AIDS] progression" that health officials halted a trial ahead of schedule because of the extremely "high rates of adverse side effects."[91] Among the "side effects" in the study, according to the *Times*, were a "more rapid rate of disease [AIDS] progression . . . and death."[92] In other words, if physical deterioration is speeded up by AZT, it proves that it causes immune deficiency disease—such as "AIDS." And since "death" is a side effect of AZT, the standard treatment for HIV and "AIDS," when an individual gets AZT, he or she gets the symptoms of "AIDS." Is it unreasonable to conclude that the prescription of AZT is iatrogenic medicine?

Sounding another death knell for AZT is a recent study by the Harvard Medical School that concluded that "some AIDS patients taking AZT who developed resistance to the drug are nearly three times more likely to die than patients not taking AZT."[93]

DAIDS Prevention and Safe Sex

To be truly effective, our nation's educational "AIDS" prevention campaign should single out drugs as immune suppressants, and it should identify drug users as the primary source of transmitting HIV and many other microorganisms. Even clean needles should be identified as dangerous, and IV-drug use should be decried as the highest risk habit, because what is in the psychoactive drugs inside the needle is far more lethal than what is on the needle itself. A warning should be issued that receptive anal sex with drug users, especially without latex condoms, is the most efficient mode of transmission for HIV infection and that it is a pathway to many of the old diseases called "AIDS." The education campaign should include truthful information on AZT. The use of latex condoms lubricated by additional Nonokynol-9 should also be strongly encouraged for protection against sexually transmitted diseases and pregnancy, as well as HIV transmission, but should not be relied on solely to prevent infection.

Another protective measure involves good hygiene. Washing the genital area and under the fingernails with soap and water before and after intercourse and avoiding digital penetration of the vaginal area are believed to be the primary reasons that infections are kept so low among non-drug-using prostitutes. This kind of hygiene and "safe sex" may also reduce the incidence of cervical cancer.

Abstinence from sexual intercourse should be put on an equal footing with the advocacy of condom usage because it more effectively achieves the objectives of condoms and strengthens character development among young people. The truth is that "AIDS" is drug-related. Therefore, I advocate the following anti-"AIDS" prevention slogan: "AIDS" is DAIDS!

CHAPTER 8

THE GANG
OF FRANKENSTEIN

There is nothing more frightening than ignorance in action.
—GOETHE

This chapter will take us into the shadowy world of scientific researchers who have no moral commitment to their work or to the protection of human life. Their research is corrupted by greed, fame, and the worship of technology. They victimize laboratory animals. They victimize humans. I call these out-of-control scientists who pose our greatest threat to public health "the Gang of Frankenstein."

Others have written that the scientists who promised to cure everything from cancer to "AIDS" through molecular biology and biotechnology have led us nowhere in their research. I disagree. These accident-prone, reckless, and im-

182

moral scientists have delivered humankind to the brink of an abyss.

In 1974, the National Academy of Sciences (NAS) sounded the first alarms over dangerous and irresponsible conduct by biomedical researchers. The NAS called upon scientists throughout the world to voluntarily halt experiments that linked animal viruses. In effect, the NAS was demanding that biomedical researchers stop playing God and gambling with the future of the human race. This official notice by the NAS to the Gang of Frankenstein alerted the public to the fact that science has become a threat to human life. But the irresponsible research practices have continued unabated, and grow more deadly with each passing day.

Beyond "AIDS"

As biomedical research has become more sophisticated, and as biotechnology has expanded its ability to create deadly new microbes, the hazards of a biological disaster have grown exponentially. Even the threats of HIV and "AIDS" are gradually being upstaged by the awesome likelihood of a global plague caused by newly manufactured and/or escaped microbes from biomedical laboratory experiments. These are man-made, monstrous microorganisms. Chimeras.

The horror is that we cannot find cures or treatments fast enough for the new causes. Some bacteria will now live only in the presence of antibiotics developed to combat known organisms. Mutated tuberculosis, herpes, Legionnaires' disease, malaria, syphilis, gonorrhea, and hepatitis are now resistant to antibiotics.

Patients are dying in hospitals from postoperative infections of wounds, while desperate doctors helplessly watch antibiotics that can no longer kill the organisms. It is difficult to determine the damage done or the deaths that may yet result from irresponsible biomedical research, but a look at what goes on in the use of nonhuman primates as laboratory animals is enough to raise serious questions about the safety and the sanity of past and current practices.

Dangerous Specimens

If you talk to people who work in primate research, you'll find they have a great fear of being bitten and infected by their research animals, and with good reason. Of all animals used in scientific research, monkeys pose the greatest risk to humans. Until the 1930s, the monkey B virus, a form of herpes, was virtually unknown to science. However, in 1932, the B virus killed a scientist after he was bitten by a laboratory monkey. The research community was troubled for years as to what caused the death of the scientist, who came to be referred to as "Dr. W.B."

The answer began to emerge nearly twenty years later, just after Dr. Albert Sabin and Dr. Jonas Salk made medical history when, working independently, they each developed a vaccine for the crippling poliomyelitis. By 1955 one million Americans had been inoculated against polio with Salk's killed-virus vaccine, but around the same time scientists began to suspect that the same deadly B virus contracted from research monkeys by Dr. W.B. in 1932 could have contaminated the polio vaccines. A live virus was found in one batch of the Salk vaccine, and this resulted in a polio epidemic among children between 1956 and 1957.

184

The Monkey Threat

All of the polio vaccines had one thing in common: They were produced from the kidneys of monkeys. The Salk killed-virus vaccines were routinely treated with formalin, a toxic solution of formaldehyde in water, to destroy the vaccine virus and the dangerous monkey B virus. But a second form of monkey virus was discovered in 1961—a papovavirus, simian virus 40 (SV 40), also found in the monkey's kidney tissue. SV 40 had survived the formaldehyde.

This harrowing discovery suggests that between 1953 and 1961, millions of Baby Boomers were inoculated with the highly infectious SV 40 contained in Salk's killed-virus shots and Sabin's live-virus sugar cubes during mass polio vaccine trials.

"It appeared that while physicians were vaccinating the American population against polio, they were simultaneously—between the years 1953 and 1961—inoculating them with SV 40. Even worse, the virus was infectious. A tentative, hasty sample of blood tests turned up many people who were antibody-positive for SV 40,"[1] writes Deborah Blum, author of *The Monkey Wars*.

Blum calls our battle against monkey viruses "a curious paradox." She writes that even as we bring monkeys out of the rain forest for the purpose of preventing human illness, we are bringing forth monkey-carried diseases that kill people.[2] Recent events give deadly weight to her words.

Attack of the Microbe

In 1976, Sudan and Zaire were attacked by a horrifying and highly infectious filovirus named Ebola. This Zaire filamentous virus strain was identical to the Marburg (Germany) outbreak in 1967, which spread to Frankfurt and Belgrade, Yugoslavia. That outbreak was traced to a shipment of African green monkeys from Uganda, which were used in polio vaccine production.

Ebola is the deadliest virus known to man. The Ebola death is quick, but the pain is excruciating. The body simply and horrifyingly liquefies. Blood clots form, damaging the internal organs, notably the liver, kidneys, and intestines. Blood flows from the eyes, mouth, ears, nose, nipples, and even cracks in the skin—every orifice of the body. In the end, the body is awash in disintegrated blood, the skin pulpy and oozing. The victim suffocates on his or her own uncontrolled bleeding.

The Unknown Threat

Can you imagine the American public being exposed to this horrifying disease? Well, it almost was. In 1989 at the Hazelton Research Primate Quarantine Center in Reston, Virginia, a suburb of Washington, D.C., research monkeys shipped from the Philippines were found to be infected with Ebola Reston, as it is now called.

After the fear of this biological meltdown passed, the center was immediately decontaminated and shut down. This horrifying scenario was quickly seized upon by the media and by Hollywood. The 1995 Dustin Hoffman movie

Outbreak, billed as a "biomedical thriller," was loosely based on the Reston incident; the film provoked so much public anxiety that the CDC in Atlanta issued a press release that at the very least was erroneous, misleading, and irresponsible.

A CDC official told the media that an airborne Ebola virus is not a reality and claimed there have been no documented cases of Ebola in the United States. Both statements are lies. Ebola Reston was transmitted through the air, and unlike most microorganisms, filoviruses will cross the species barrier from monkey to man—and back to monkey.[3] Also, one animal handler, according to Blum, tested positive for Ebola Reston, and Laurie Garrett reports in *The Coming Plague* that five animal handlers at JFK airport in New York have been infected with Ebola Reston. A New York television station reported an Ebola-infected White woman who is an animal handler at a New York City airport. The CDC official also assured the world that "the potential for introduction of Ebola outside Zaire is low."[4] These reassuring words only frighten you further when held up to an assessment of the U.S. government's handling of the Ebola outbreak in Reston as provided by the U.S. editor of *Lancet*. Writing in *The New York Review of Books*, Dr. Richard Horton noted that "the [U.S.] military instituted a planned policy of disinformation even though two men had already been taken ill." The government's intent, he said, "was to mislead." Horton wrote that "to stave off unwanted attention from the mass media, children were allowed to play freely around the Reston unit."[5]

Contrary to the CDC's disinformation, the only reason that you are reading this book is because you and I are lucky—lucky that the Ebola Reston virus and the Ebola strain that appeared in Pennsylvania a month earlier at a quarantine center, unlike the 1976 Ebola Zaire or the 1995 Ebola Kikwit, were not the most lethal Ebola strains.

But escape from Ebola Reston by no means gets us out of harm's way of a new viral assault on humans. In fact, if you are one of the millions of American citizens who was inoculated against polio some thirty years ago, there may be a virus running through your veins that could be even more dangerous than HIV. It is the SV 40 virus, simian virus 40.

A Known Threat

In *Some Call It AIDS, I Call It Murder*, Dr. Eva Lee Snead makes the case that laboratory monkeys transmitted the "AIDS" virus to humans through contaminated polio vaccines. She writes that the viruses—including SV 40—that contaminated the polio serum were "slow viruses" that produce an extended symptom-free incubation period in monkeys. Therefore, the polio vaccine contamination provided a portal for the "transfer of slow and difficult-to-detect diseases" from animal primates into the human race.[6]

As Sabin and Salk were working during the 1950s to develop a vaccine, another doctor, Hilary Koprowski, was also laboring to find a polio cure. Salk was the first to become successful when, in 1954, he introduced a "dead" poliovirus that caused the body's immune system to manufacture disease-fighting antibodies effective against the paralyzing forms of polio. However, it was a weakened live-virus polio vaccine developed by Sabin that became the vaccine most widely used. Sabin's was an oral vaccine taken in a sugar cube. It required no injection or booster shot as did

Salk's, and, if not contaminated, it frequently gave a lifetime immunity from polio. Salk believed that on rare occasions Sabin's live-virus polio vaccine produced polio instead of immunity.

Koprowski also introduced an oral vaccine, and he competed with Sabin to mass-market his version. Unlike Sabin's sugar cube, Koprowski's vaccine was sprayed into the mouth.

In 1957, Koprowski left for Africa. In six weeks' time, nearly one-quarter of a million Africans had been inoculated.

The First "AIDS" Case?

The Koprowski polio vaccine was given in 1957 in Léopold-ville in the Belgian Congo (today it is called Kinshasa, Zaire), which is also where the first case of "AIDS" appeared. According to a *Rolling Stone* article on the origins of "AIDS" by writer Tom Curtis, many scientists believe that the "AIDS" syndrome "dates from a plasma sample drawn in 1959" in Léopoldville.[7] Another article—this in the medical journal *Lancet*—confirms that the earliest type B HIV-positive blood sample was taken in 1959 in Léopoldville.[8] If that is true, what could have been a vaccine-related "AIDS" appeared in Africa about twenty years before "DAIDS" did in the United States.

Curtis explained the connection in the *Rolling Stone* article: "The world's first mass trials of an oral polio vaccine . . . took place from 1957 to 1960 right in the middle of what was then the Belgian Congo, Rwanda and Burundi—the epicenter of the future African AIDS epidemic."[9] A detailed map drawn by Koprowski himself in 1958 shows

189

where he gave his inoculations in the northeastern area of the Belgian Congo; today this is "the region of the highest HIV infection in equatorial Africa," Curtis reported.[10]

Curtis's theory, however, would later be challenged. His perceived attacks on Koprowksi led to the formation of an independent research team—called together by the Wistar Institute, one of the original manufactures of polio vaccines and Koprowski's former employer—to examine the allegation. The team's findings, which were published first in the *New York Times*, said that Curtis's theory was not only "highly improbable," but also "so unlikely that we don't have to take it seriously."[11]

Citing a patient named John Carr as the first documented "AIDS" case, the Wistar group concluded that the chances of Curtis's theory being valid were "extremely low" if "AIDS" appeared in Africa in 1959 and Carr died of it in England the same year. Carr, a sailor who neither was ever in the Congo nor had any contact with Koprowski's polio vaccine, allegedly died of an "AIDS"-related disease in 1959 in Manchester, England. Conspiciously, the *Times* account neither made mention of nor gave credence to the case upon which Curtis built his theory: the 1959 Léopoldville strain. Neither was there a unanimous consensus among the committee members that Curtis was wrong. Dr. Frank Lilly of the Albert Einstein College of Medicine believed that "it was actually a plausible hypothesis."[12]

In its December 9, 1993, issue, however, *Rolling Stone* published a "clarification" of Curtis's original story stating that its editors never intended to give the impression that it was based on scientific proof, nor did they know of anything that would suggest that Dr. Koprowski, because of any careless procedure, was responsible for the creation of "AIDS."

Curtis, while agreeing that his theory could be flawed, still contends that because of the tremendous number of people inoculated, if only a small number had sores in their mouths or inhaled the virus deep into their lungs, year-to-year transmission could have produced the unprecedented number of HIV infections (6.5 million) that we see in Africa today.

In another instance, both the disputed Carr case and the "Léopoldville strain" of HIV in 1959 have also been cited by those seeking to refute the widespread conspiracy theory that "the AIDS virus" was purposefully created in a biomedical laboratory. The sophisticated technology did not exist in 1977 that would have allowed retrovirologists at Fort Detrick, Maryland, or anywhere else to genetically engineer a synthetic virus from bits of genetic material from bovine leukemia virus, sheep visna virus, horse and goat lentiviruses, and HTLV-1, as Dr. John Seale alleged in the *Journal of the Royal Society of Medicine.*[13] There is also the matter of the disputed appearances of HIV and "AIDS" simultaneously in Manchester and Léopoldville in 1959, which suggest that the manufacture of a synthetic virus was unnecessary had the government wanted such a biological weapon.

However, it is possible that SV 40, a "slow virus" with an extended gestation period, which we know contaminated the polio vaccines of the 1950s, provided a portal and subsequent sanctuary for other "slow viruses" that scientists some twenty years later would call HIV or HHV-6, which is thought by some researchers to be associated with chronic fatigue syndrome. HHV-6 is a strong contender as a cofactor in "AIDS." The research establishment has known since 1961 that a contagion of forty simian viruses had the potential of unleashing a spate of parasites, such as the toxoplasmosis that Arthur Ashe got in a blood transfusion, and

Pneumocystis carinii—one of the major causes of death by pneumonia in "AIDS" cases.

Moreover, there is no proof that the polio vaccine used by Dr. Koprowski was in fact responsible for introducing "AIDS" into the human population, and Dr. Koprowski denies that "AIDS" could have been communicated by the Congo vaccine. Nonetheless, the coincidence that these mass polio vaccine trials occurred in what later became the "epicenter" of "AIDS" in Africa raises interesting questions.

The Manchester case has been recently challenged. Doubts were raised when a reanalysis of the two samples by Dr. David Ho suggested that the two sets of clinical materials came from more than one person. In response to Ho, a University of Manchester research team stood by their 1990 findings.[14] On the other hand, according to Garrett, the "Léopoldville strain" was never confirmed by polymerase chain reaction (PCR) data because it was lost by American scientists.[15]

Despite this confusion, the potential appearance of two cases of "AIDS" thousands of miles apart almost simultaneously could be cited to make the argument that each evolved independently. Laboratory experiments with human leukemic blood and with viruses from several monkey species were being conducted during that period in England as well as the United States. Those experiments could somehow be related to the Carr case.

The contaminated American-made SV 40 polio vaccine used in Africa in the late fifties may also offer clues to explain why a variant of the European–North American HIV type B appeared there twenty years before it did in the United States. Another indication that there may be several different natural sources of the HIV is offered by the fact that genetic typing of HIV confirms that six different groups of the virus exist. Type A HIV is found in "AIDS" patients in Central Africa and India, and type B HIV is associated with "AIDS" cases found in the Western Hemisphere.

Another significant link between vaccines and the history of "AIDS" is the fact that shortly after the Belgian Congo received its independence and became Zaire in 1960, the exodus of Belgian colonists left a huge void in the work force. That void was filled largely by French-speaking Haitians living in Zaire. Could there be a link between the presence of these Haitians in Africa and the incidence of HIV among Haitians today?

The Coming Cancer Epidemic

There is a twenty-five- to thirty-year incubation period for the slow-acting viruses such as SV 40 that contaminated the polio vaccines given to thirty million Americans between 1953 and 1961. It has been speculated that the presence of SV 40 in the early polio vaccine explains why the incidence of cancer increases each year. In a study in which SV 40 was injected into mice, the mice developed Kaposi's sarcoma–like tumors. Whatever the cause, the spread of cancer continues in spite of annual increases in funding for cancer research. In testimony before the U.S. Congress, Dr. Paul Calabresi, chairman of The National Cancer Advisory Board, predicted that cancer will become the nation's top killer by the year 2000.

Another thought to ponder. Perhaps what this country needs is not an "AIDS" test, but a drive to detect the presence of SV 40 antibodies. Dr. Edwin H. Lennete, a viral expert, reports "high and persistent antibody levels"[16] to the SV 40 virus in people who received polio vaccines. Dr. William F. Koch, author and scientist, goes so far as to present statistical evidence to support his thesis that "paralytic polio is increased both in incidence and fatality by use of the vaccine." Koch pointed out that "in North Carolina and Tennessee, where vaccination was compulsory, there was a 400 percent increase."[17]

Animal Origins

In a 1987 report by the National Antivivisection Society entitled *Biohazard: The Silent Threat from Biomedical Research and the Creation of AIDS*, the thesis was presented that "given the nature of experiments during the 60s and 70s, AIDS more than likely was created by scientists, whether knowingly or unknowingly, in a laboratory."[18]

Biohazard pinpoints the beginning of a biological chain reaction in the late 1960s, when there were several outbreaks of cancer in laboratories around the world that housed and experimented on animal primates. Some of the sources of these outbreaks were traced back to human leukemic blood that had been injected into a series of gibbon apes. Nine of the gibbons developed leukemia and all of them had a strain that came from the same two human blood sources. After a succession of experiments, one of which included injecting bone marrow from the leukemia-infected gibbons, a type C virus developed in two gibbons.

These experiments continued, and by 1969, this new type C virus was causing unprecedented cases of cancer in the animals at the California Primate Research Center at

Davis. Many of the animals died of "AIDS"-like illnesses. By 1971, the researchers were well aware that the virus could infect human tissue. They also knew this experimentally induced disease produced symptoms similar to those seen in homosexual men at that time and that it could kill humans. That knowledge did not slow their deadly science.

Virologists and vivisectionists are repeating experiments of this type every day, taking a killer virus from one animal and introducing it into another—mixing them and then injecting this unknown combination into human cancer cells. The results are frightening microbial monsters, Frankenstein-like horrors that nature is not capable of conceiving. The scientific world was warned as far back as 1966 about producing these new pathogens. Laurie Garrett, author of *The Coming Plague,* writes that officials at the CDC and the World Health Organization admit that they are "woefully lacking"[19] when it comes to dealing with the potential disaster posed by microbe bombs. Many scientists, including Nobel laureate Joshua Lederberg, believe that now, more than any other time in human history, the world's population is vulnerable to global disaster from a viral epidemic.

Scientists are well aware that a virus may be forced to mutate and that it may become more dangerous when it is transferred across the species barrier from its host animal to another species. Bringing together and gang-caging diverse species of animals in laboratories and repeatedly introducing alien organisms into their bodies has produced biological time bombs.

Microbe Menu

To understand the cataclysmic potential of this Franken-
steinian practice, consider this series of laboratory experi-
ments that began in 1980 at the New England Regional
Primate Research Center in Southboro, Massachusetts.[20]
First, blood from cancerous tissue in a monkey (1) was
injected into two other monkeys (2 and 3), both of which
died of cancer. In a separate experiment at the same lab,
infected tissue blood from another monkey (4) with a differ-
ent cancer was injected into two monkeys (5 and 6), which
also died. All four of the injected monkeys (2, 3, 5, and 6) died
as a result of a newly created immunosuppressive virus that
was first identified in monkey 6. Tissue from monkey 6 was
then injected into seven other monkeys, all of which died
from infections due to immune disorder.

Blood from three of the seven dead monkeys was then
mixed and injected into another six monkeys, all of which
died within sixty-five days. Blood from these six monkeys
was then mixed and further injected into two other mon-
keys; they also died. A new lethal virus had mutated and
was found in the last two monkeys; they both died from
brain damage—a condition often associated with human
"AIDS."

Another particularly frightening aspect of these danger-
ous experiments on animals arises from the exorbitant cost
of lab monkeys. Chimpanzees, for example, now cost as
much as $60,000 each, and there is a growing black market
for them. Monkeys that should be destroyed are instead
being resold after experimentation. These sick monkeys car-
rying infectious and unknown microbes are being traded on
the international market, and it is impossible to know what
diseases they carry—as in the case of Marburg and Reston.

MONKEYS ARE A MICROBIOLOGICAL HAZARD AND RECENTLY
CAUGHT ANIMALS ARE AN UNRELIABLE RESEARCH TOOL, read an
advertisement from Shamrock Farms, a primate supplier in
Great Britain.[21] It also warned that for-sale monkeys were
likely to carry tuberculosis, bacteria, parasites, microorgan-
isms in the blood, and so on. There is a distinct possibility
that monkeys may be a source of newly mutated old micro-
organisms or newly created laboratory microbe bombs.

Is There an "AIDS" Conspiracy?

The best evidence that there is a conspiracy to extermi-
nate Blacks as well as homosexual men both in Africa and
in the United States is the disproportionate number of HIV
and "AIDS" victims among these groups. But I believe this
fact is more likely due to high-risk drug use and risky sexual
behavior among small segments of these two groups. Even
if a genocidal plot exists to kill off Blacks, Latinos, and ho-
mosexuals with HIV and "AIDS," it cannot succeed without
the cooperation of these groups through drug usage. It has
been suggested that HIV is benign unless your body's im-
mune system is chemically poisoned and/or malnour-
ished—"thirdworldized."

Increasingly there are reports of HIV-positive individuals
who are healthy ten years after their initial diagnoses. New
Yorker Salvatore Fortunado has reportedly carried the HIV
virus for twenty-five years without experiencing any cell
damage. I challenge the medical establishment to find a
twenty-five-year survivor of AZT—or any other current
chemical treatment.

Ethnic Weapons

If racist conspirators or rogue agentry wanted to kill off Blacks, why not use one of the chemical and biological "ethnic weapons" the United States government has already developed? These racially specific viruses target only individuals with the approximation of an African genetic coding or any specific population group for incapacitation or death while doing minimal damage to friendly forces.[22]

The chief of a Soviet news agency reported in 1987 that the CIA had "a biological weapon for killing Blacks," according to a story in the *Chicago Tribune*.[23] In 1989, biotech expert Raymond A. Zilinskas warned the American Association for the Advancement of Science that before the year 2000, biotechnology will perfect race-specific biological weapons.[24] Advances in science and technology have not simply made killing easier on a mass level, but more selective.

Yes, researchers have developed biological agents and organisms that target specific racial groups. Instead of "designer drugs," one U.S. Deputy Secretary of Defense allegedly boasted to a congressional committee about "designer BW"[25]—we now have biological warfare that is genetically engineered. *Newsweek* magazine explained the practical application for studying racial differences. Genetically engineered, self-destructing microbes could make "germ warfare more precise."[26]

Newsweek reported: "It might even be possible to wage ethnic warfare by developing substances that affect one race more than another; examples include valley fever, which is much more likely to kill blacks than whites."[27] A naval bioscience experiment revealed that Blacks are ten times more likely to die from valley fever than Whites because of melanin skin pigmentation.[28] Other biological warfare favorites include Lassa fever virus, anthrax toxins, botulism, and Ebola virus.

Specific Targets

Semites. In 1975, Dr. Richard Hammerschlag, a neuroscientist, revealed that three years earlier the Pentagon had sponsored an elaborate study of a "complex chemical vulnerability in Semitic peoples." Hammerschlag estimates that, with an ethnic weapon, 80 percent of a Semitic army— Arabs or Jews—could be destroyed.[29] It has been rumored that Hammerschlag refused to reveal the necessary chemicals for fear that terrorists might exploit the information. But believe me, if there is a formula, it will one day fall into the wrong hands.

A testimonial to that chilling assertion is the discovery of Iraq's major germ warfare program with enough of the bacterium *Clostridium botulinum* and *Bacillus anthracis* to kill billions of people. This makes the fact that Iraq bombed Israel in 1991 even more ominous.[30]

Blacks. In 1951, U.S. Army scientists wanted to find out if a lethal fungus, *Aspergillus fumigatus*, would specifically attack Black people. The Army scientists intentionally contaminated the crates to be handled by Black workers on the docks at the Norfolk Supply Center in Virginia. The Army used coccidioides, a fungus disease derived from *Aspergillus fumigatus* mutant C-2, to which Blacks are genetically more susceptible than Whites. Unsuspecting Blacks were exposed to a fungus at the Mechanicsburg, Pennsylvania, naval depot because "Negroes are more susceptible to coccidioides than are Whites," a Pentagon official told a Senate subcommittee when testifying about biological testing on human subjects in 1977.[31]

The same "ethnic weapons" research was allegedly conducted from 1956 to 1958 on Blacks in Savannah, Georgia, and Avon Park, Florida. The army released mosquitoes that were possibly yellow-fever-infected into the Black residential areas of these communities from ground level and from airplanes and helicopters, according to Robert Lederer in the *Covert Action Quarterly*. He reported that many people were swarmed by mosquitoes and that they may have been infected; some died. After each assault, Army agents posing as public health officials would photograph and test their victims. They then fled the Blacks and their dangerous fevers.[32]

Targeting Blood Types

Ethnic weapons constitute a new class of chemical and biological warfare (CBW) agents that can exploit natural differences in gene frequencies among specific population groups. The seminal work on this subject is an article that appeared in the *Military Review*'s November 1970 issue; simply called "Ethnic Weapons," it is by Dr. Carl A. Larson, who was the head of a university genetics department in Sweden in the 1970s. In technical terms, he outlined the "polymorphisms" or blood types found in different frequencies among various sociological population groups.[33]

Larson implies that scientists have already developed biological agents to exploit the blood types (A, B, AB, and O, as well as the Rh factor). For example, 13 to 17 percent of the White stock of Europe, America, and Australia is Rh-negative; American Indians "are in the fortunate position of having no Rh-negative genes."[34] The incidence of Rh-negative genes among Black Americans is about half that among White Americans, indicating the genetically hybrid origin—the African, European, and American Indian stock—of "Black" Americans. On the other hand, blood type B never occurs in American Indians, but in the people of Southeast Asia and southern India it occurs 30 to 40 percent of the time.

Larson was excited about the fact that a special enzyme determines milk tolerance among racial groups. He found that only European-descended adults can drink large amounts of milk. In adulthood, many Africans, American Indians, and Asians lose the appropriate enzyme and become ill or die from milk. The white-skinned Europeans retained the enzyme because it evolved when the body found a way to absorb sunlight (for vitamin D) during the short winter days in the northern regions of Europe. Among skeletal remains in Europe, bowed legs due to a vitamin deficiency are common. The skin is white in the first place because of a chemical adaptation to a region of the world without long days and strong ultraviolet rays of the sun, which darken the skin and produce melanin as a sun shield.

Pandora's Box

With these biological weapons being developed, the human race may literally be on the brink of opening Pandora's Box.

Perhaps we have already peeked into it but have not realized what we were looking at.

Environmental factors, including years of experimental vaccines, malnutrition, poor hygiene, and a lack of medical services, have already predisposed the Third World to microbal assaults. As benignly as "the AIDS virus" (HIV) behaves in healthy hosts, its potential for havoc is awesome, as we see when it finds susceptible hosts; even in the Western Hemisphere, we now see the phenomenon of "thirdworldization" among high-risk groups.

The doomsday infrastructure is already in place. Biological disaster could result from a planned assault on a specific population target, or from an accident in a biomedical research lab where far too often moral and safety concerns are secondary to the pressures of ego and economics.

An Ebola microbe bomb could have exploded when it escaped in Reston in 1989. At Yale University in 1994, a laboratory researcher did not report a serious exposure to the deadly Brazilian Sabia virus. He did not quarantine himself as required and, without regard for human life, went out into the community and recklessly exposed a family, including two young girls.[35] Our luck will eventually run out. It will take only one accident to precipitate a biological endtimes.

I believe there is a 100 percent chance for an epidemic. Biomedicine has become corrupted by egomaniacs and the rush for economic gain. Scientists create biological monsters by introducing viruses from one species into another. And they ship disease-ridden animals all over the world. These viruses, bacteria, and protozoa—microbe bombs—are gradually finding their way into the human population. I seriously doubt the latest establishment theory that Ebola and other new mysterious microorganims are nature's punishment for environmental crimes, just as I reject the mainstream myth that HIV equals "AIDS."

Science Without Virtue

I do not wish to leave the impression that I believe all scientists, public health officials, and community leaders are untrustworthy. In fact, many, if not most, are unsung heroes. Nor do I wish to suggest that the U.S. government or any other government is conspiring against its own people—although I suspect that a handful of rogues do control sensitive areas within most governments in the world. However, retired Colonel L. Fletcher Prouty, who served in the Pentagon as the Department of Defense liaison to the CIA, reports in his book, *The Secret Team*, on a rogue "High Cabal" in the CIA that has tentacles throughout the government and business.[36] Other governmental rogue units may exist also at the National Institutes of Health, where some are peddling the AZT snake oil; and at the Food and Drug Administration, which approved dangerous experimental drugs and vaccines to be used on four hundred thousand troops during the 1991 Persian Gulf War. A Senate report confirmed that the soldiers were not warned about the adverse effects of the FDA-approved pyridostigmine bromide, among other experimental drugs, nor were they told that the Pentagon knew the risks before ordering them to use the drug.[37]

This unconscionable behavior and the plethora of diseases that resulted among the Gulf troops and their families—the Gulf War Syndrome—are the latest instances in a fifty-year history of our government's apparent disregard for the rights and lives of its citizens. Government-sponsored gas chamber experiments, exposure to radiation, LSD experimentation, clandestine "body snatching" of thousands of poor urban Blacks in the 1950s to study the extent of radioactive fallout[38]—all of it is science run amok.

And now genetically engineered synthetic viruses and racial bombs. Moral rationalism reduces our world to a survival-of-the-fittest jungle and produces a modern-emotivist climate in which taxpayer-supported Frankenstein scientists and rogue bureaucrats have become the angels of death.

CHAPTER 9

TEAM AMERICA: THE END OF RACISM AND SEXISM

If we Americans are to survive it will have to be because we choose and elect and defend to be first of all Americans; to present to the world one homogeneous and unbroken front, whether of white Americans or black ones or purple or blue or green.
—WILLIAM FAULKNER

After World War II, the United States emerged as a super-power with proven industrial might. But shortly thereafter, most of the country headed for the golf course or the old fishing hole. We lost our competitive edge, and as a result, our economic growth slowed to a crawl. The United States came out of that war with incredible advantages. Every bombed-out factory in Berlin added to the value of those in Detroit. We were ahead of the other industrialized nations financially as well as in food production, technology, and, most important, human capital—education and work experience. We had the world in the palm of our hand.

The people of the nations that we defeated did not accept defeat, however. The Germans and the Japanese accepted change and sacrifice instead. They restructured their societies and reorganized their industries. What did we Americans do? We wasted the grand opportunity to secure our place as the world's greatest superpower.

Instead of building a nation that could lead the rest of the world into the future of technology, we took early retirement. Rather than building on the strength that grew out of the war, we rested on our laurels. We convinced ourselves that all we had to do was to put "Made in U.S.A." on a product and the rest of the world would snap it up. And in no time at all the label "Made in Japan" went from a joke to a mark of excellence.

While our international competition invested in the future through change and sacrifice, we lived for the moment. We bought cars on credit. We mortgaged and second-mortgaged homes and farms. We found ways not to create wealth but to produce debt and to encourage greed and fiscal irresponsibility. We literally created junk bonds that produce nothing, and with them, a junk economy. As a result, we fell into debt as individuals and as a nation.

Once considered an invincible economic superpower, the United States today is a global grub looking for handouts from nations that previously came begging to us. The world's greatest lender is now its biggest debtor. We are at the mercy of foreign investors and speculators who attack the American dollar at will and indirectly dictate economic policy to our impoverished leadership. If we had to go to war, we couldn't because we couldn't afford the tab. Like a heavily armed tank flipped over, this country is virtually helpless. Debt has upended us. We have the resources, but because we consume so much more than we produce we must go begging to foreign nations for assistance.

Ironically, we may be in Peter Drucker's *Post-Capitalist Society* in which knowledge is more valuable than capital. The competitive advantage in a high-tech globalized economy is now increasingly determined by human resources, or human capital—the form of wealth that America has ignored and abused the most. And it has caught up with us. The primary reason that we are in a productivity decline is that our workers are increasingly not as "smart." We are increasingly unskilled and untrained, especially when compared with the Japanese and the Germans. We are losing the battle primarily because of our underdeveloped human capital.

Where does America truly rank in a high-tech, globalized economy? Our school systems are no longer solely dedicated to education, discipline, and imparting fundamental knowledge. Instead, they have become battlegrounds, not only for gang warfare, but also for special interests and social experiments—everything from forced integration to sex education, gay pride, and the tug of war between church and state. Adults play their power games while the children struggle to read what is printed on the back of the cereal box. A Carnegie Foundation study reported last year that American high school graduates are so poorly prepared in math and quantitative reasoning skills that they ranked last in a study of fourteen countries.[1]

As usual, the failures of our nation are most obvious among the most neglected: our urban and rural poor, primarily the Black underclass. Other than grudging concessions to civil rights pressure, no enduring efforts have been made to halt the marginalization of Blacks. Racism causes economic inefficiencies and societal disorder. It creates an oppressed class that incubates societal pathology and, like a cancer, spreads economic and social breakdown. Because of technological displacement, the American middle class is in danger of disappearing. And without a strong middle class, a nation takes on the status of a Third World country restricted to only the haves and the have-nots—with each dependent on government handouts. Economic, social, and moral decline renders a racist nation unable to compete with nonracist countries that are not crippled by the same internal diseases.

Will It End in a Bang or a Boom?

This chapter is about ending the decline of the United States of America. It is about ridding this nation of the crippling debt and the twin shackles of entitlement socialism and racism. It is about encouraging Americans to embrace two foreign, but liberating, concepts—sacrifice and change.

From all corners and in every way, America is being encouraged to reinvent itself in order to survive. And if we are smart, we will develop a plan and a learning curve for shared sacrifice and change. This is the agenda that worked for the Germans and the Japanese, and it can work for the United States.

I call this new vision *Team America*, because the concept of a team implicitly suggests cooperation—and sacrifice. As we enter the next millennium, we will become a team whether we like it or not. Or we will continue to struggle for survival as a bickering collection of self-serving special interests that is generally irrelevant in the global scheme.

According to the Hudson Institute, by the year 2000, the U.S. workforce will be significantly female and non-White (female 47.5 percent; non-white 15.5 percent). At the same time, the White male birthrate is declining.[2] This singular demographic fact has mandated the inclusion of groups traditionally excluded from the labor pool. America will change not because you or I demand it. It will change because soon, Americans will have no other choice. Those who are qualified but have traditionally been shut out will be sought because the color and gender of the labor pool has changed because of market demand.

In the next five years, the percentage of Whites and males will diminish further. That means that the children of today's White Americans will live in a country dependent, in large part, on the skills and productivity of non-White Americans and women. Cultural diversity is rapidly becoming an inescapable reality. What does this change in workforce demographics mean? It means we need the best-qualified person in the workforce regardless of race, ethnicity, or gender. Because of marketplace demands the traditional White male workplace preference will go the way of affirmative-action preference programs for women and non-Whites. Take a look in the mirror and greet a member of the new, improved Rainbow Coalition.

Somewhere over the Rainbow

Twenty-five percent of Americans now define themselves as Hispanic or non-White. If current trends in immigration and birthrates continue, by the end of the century, the numbers will look like this: The Hispanic populations will grow by an estimated 21 percent; the Asian presence by about 22 percent; the Black population by 12 percent, and the White population by only 2 percent.[3]

By 2020, the number of non-White U.S. residents will have more than doubled to nearly 115 million, while the

White population will not have increased at all. If the racial makeup of America continues to change along these trend lines, according to the U.S. Census statistical definition, the average U.S. resident will trace his or her descent to Africa, Asia, the Latin countries, the Pacific Islands, or Arabia.[4] This phenomenon opens an unprecedented window of opportunity for women and non-Whites, as well as any America with twenty-first-century skills. And it will not happen at the expense of White males, not if they join the team rather than isolating themselves. For the nation, as a whole, Team America presents an opportunity to restructure as an egalitarian society—free of gender, racial, or ethnic institutional bias—and as a competitive player in a globalized economy. These seemingly utopian goals are attainable if we follow a vision of Team America and adopt cultural diversity as our country's economic salvation. The alternative is a national breakdown similar to the advanced socialism that destroyed the Soviet Union.

The key to cultural diversity and survival as Team America is how we manage our human resources—the people who make up the most culturally diverse workforce in our nation's history and, perhaps, in the history of the world. In this context, getting along with one another takes on an entirely different meaning. It now means getting along with someone else and getting paid for it, without government coercion. In the coming decades, the members of your team may not look like you, but how well you work together may well determine your standard of living—the best incentive for gender and racial peace we have ever had.

The new global game of the twenty-first century will pit similar economic systems, with the most efficient coming out on top. It will be capitalism versus capitalism as opposed to capitalism versus Communism, which was the game of the twentieth century, according to Lester Thurow, dean of MIT's Sloan School of Management and author of *Head to Head*. Thurow has projected that we have about five years to get ourselves organized to play that game.[5] The Asian, European, and American people will be competing head to head. How do these teams match up? One big question is whether the American automobile industry can survive another twenty years of Japanese competition.

World Cup Economic Competitors

In spite of a few recent economic setbacks, Japan Inc. looks to be an extremely strong player. Intact, rich in financial capital, and deep in human capital, with perhaps the best educational system in the world, Japan promises to be a dominant force. The former West Germany is absorbing the economically and socially weaker East German population. As a result, in ten to twenty years, the united Germany may rank either first or second as a global industrial power. If trends continue, Japan and the Big Seven nations of Europe will catch up and, just after the turn of the century, surpass the United States in productivity, according to economic forecasts.

Faced with such keen global competition, the United States must pull together internally. As this nation now stands, it is divided racially and, increasingly, culturally. Millions are unemployed not because they won't work but because the U.S. economy must maintain a fixed unemployment rate and a growth ceiling in order to control inflation caused by the insidious national debt. The American workforce has been transformed from high-earning steelmakers to minumum-wage burger-flippers.

If the current trends continue, the United States will be lucky to remain a competitor for third behind Japan and Germany. In spite of this, our leaders act as though there were no threat to our global standing. Where is the serious effort to restructure education to produce technicians and smart workers for a high-tech globalized economy? The nation's wealth is being squandered on consumption, and there appears to be little serious interest in legislating to encourage saving and investment for economic growth. Instead, the current emphasis on entitlement socialism has enlarged the benefits giveaway to the middle and upper classes. Along with the mania of wealth redistribution through federal benefits and taxes, there is no other plan and no vision and seemingly no will to turn America around, other than the new GOP Congress, which, on its own, will have only a limited impact. It seems obvious that we can compete only as Team America. But there is no movement in that direction. Even the new Republican revolution has missed this point.

The Numbers Crunch

Sometimes I wish that we were mere numbers so that we could look at the problem of racism and divisiveness unemotionally. Numbers do not lie, nor do they discriminate. John von Neumann, who was widely believed to be the premier intellect in an era of Albert Einstein and Kurt Gödel, reduced human behavior to numbers in what has come to be known as "game theory." Although he applied game theory to strategic nuclear planning at the Rand Corporation, von Neumann also used it in an attempt to unravel human paradoxes.

The most famous of his games is the Prisoner's Dilemma. Its usefulness here is to understand group conflict and to help us identify the major stumbling blocks to cooperation by diverse people. In von Neumann's mathematical dilemma, two men are arrested and held in separate cells. They are both guilty of the same crime. Each is given the choice of turning state's evidence and accusing the other. If only one accepts the deal, he goes free and the other stays in jail. If each implicates the other, each serves two years. If both remain silent, each has to serve only one year; but in so doing, each runs the risk that he will be betrayed by the other.

What should each prisoner do? If each adheres to rational self-interest, he could harm himself, because both can use the same negative strategy. However, if they cooperate with each other, both will win.

For all of von Neumann's emphasis on numbers, the bottom line here is faith in someone else. In the Prisoner's Dilemma, as in the self-defeating conflict among America's diverse special interests, "the common good" is thwarted by "individual rationality."[6] In our country, individual rationality is manifested in the "where's mine?" attitude. It is also present in the nonstop contest for racial dominance.

You might say that this dilemma of subverting the long-term greater good for short-term individual gain confirms the old adage "Common sense is not common." The Prisoner's Dilemma is paralleled in the conflict of interests between Blacks and Whites, men and women, or any other two interest groups vying in the United States. Both players in the

Prisoner's Dilemma need the other's cooperation. Yet each player has an equally strong need to defect, meaning to not cooperate with the other. In this dilemma, as in the real-life social dilemma between Blacks and Whites—who are historical enemies because of legal slavery and majority dominance—there is no shortage of reasons why one group should not cooperate with the other.

Far too often in this country dominated by self-interest, important issues relating to jobs, wealth, education, and legislation are played out as zero-sum games in which one group's gain is another group's loss. It's odd that we have finally come to see the earth's environment as a holistic "ecosystem" in which all things are interrelated, but we cannot see ourselves that way. The results of this subversion by "individual rationality" are apparent for the world to see. The United States has no unity. It is a shattered, declining civilization with no long-term winner in sight.

As individuals, we can only develop with cooperation, help, and guidance from others. I believe that my modest individual successes have come to me personally because I sought education as a means of enhancing my natural abilities. My education came in large part from the investment and sacrifice of others—my family and even my teachers, who might personally have benefited more if they had pursued other careers.

Without an education I would not have escaped poverty and I never would have been able to overcome racial barriers to my success. Another crucial factor is that the virtues that I observed and learned from my family and my social environment enabled me to appreciate an education so that I would apply myself. As a result, I have what is called "human capital"—an accumulation of formal and informal learning, education, and training that gives me value as a citizen.

If you build human capital and use social capital, then financial capital in the form of money is much more attainable, if not almost inevitable. The key to anyone's success—whether Black, White, polka-dot, male or female—is training and education. How is America to compete when its public schools inhabit the basement of the industrialized world? One does not need to solve von Neumann's dilemma to recognize America's dilemma.

We can no longer afford to exclude capable people from our workforce. It is now an economic disadvantage to ignore Blacks and Hispanics in the inner cities when looking for human capital. "The simple fact of the matter is, they're on my team,"[7] economist Thurow said.

Same Strokes for Different Folks

Instead of using our national diversity as an asset to forge a common dynamic push for excellence, we have a vacuum. Instead of visionary leaders, I'm afraid our nation is led by Chicken Littles from the political left and right. In effect, these ethnocentrists lobby for the dominance of European culture over all others in a world that is approximately 80 percent non-White. Even Europe itself is becoming less Caucasian. As the saying goes, that dog won't hunt.

The same can be said of the diversity management consultants who have become the false gods of American business with their warmed-over affirmative-action and equal-opportunity pitches. Racism is their standard explanation for non-White retention and advancement, because it follows the logic of affirmative action. The truth is that efficient management will eliminate the impact of personal racism. Racism will be viewed as an economic liability in a truly culturally diverse environment—if goals are clearly articulated, and incentives and penalties are linked to performance and corporate culture.

Simply put, various team members will underperform in a climate where inefficient management allows intragroup conflict to act as a drag on productivity. As a result, everyone on the team is financially punished. The primary roadblocks to this process are usually White males in middle management who fear that cultural diversity is a disguised form of a zero-sum-based affirmative action, Blacks who are fearful of losing affirmative action, and consultants who know very little about the problems and opportunities created by a nontraditional workforce.

Developing and managing a diverse workforce is not simply an option for American companies. It is a task they must inevitably confront. If one company fails at it and its competition succeeds, the homogeneous company will lose market share or become insolvent because it has severely restricted its universe of human capital. As I told a group of CEOs in Grand Rapids, Michigan, if their competition is more successful in recruiting, training, and motivating skilled women, Blacks, Hispanics, and other ethnic groups that have historically not been in the workforce, then their competition will win.

White males in middle management who oppose other groups out of fear of losing their jobs should realize that their jobs will be lost unless they embrace cultural diversity, because without a diverse workforce, their companies will not be competitive. For multinational corporations, the participation of indigenous members of various countries in the management structure will be a key factor to penetrating the offshore markets where the bulk of their growth will come. The fall of the Iron Curtain and the opening of China have made the international market—with its great diversity—the fastest-growing.

In the United States, increased diversity will create powerful niche markets. New technology will efficiently target segmented markets such as gays, Blacks, women, Hispanics, Asians, etc. A more segmented society will force corporations to include more diverse representation on their staffs and boards as a sign of community commitment and goodwill. Corporate and customer relations will be transformed by the emergence of organized niche markets. Goodwill earned by managers for acknowledging cultural differences will become an increasingly potent market advantage and a corporation's first line of defense against demagogues who use boycotts as personal fund-raisers.

Cultural Diversity Is Good Business

Allstate Insurance Company serves as an example of a business that came to realize the value of cultural diversity. And that realization came out of economic necessity, not social conscience. "What is striking about Allstate's affirmative action is that it wasn't forced by law but by the company's search for new customers in a highly competitive industry,"[8] noted *Newsweek* in a 1995 profile of the $2 billion Illinois-based insurance company. *Newsweek*, however, confused affirmative action with survival in a culturally diverse society.

As late as the mid-1970s, Allstate was still a company dominated by White males. That changed rapidly when Allstate executives were confronted with saturation sales in the White market. In search of new markets, Allstate discovered the Black middle and upper classes. Its executives also realized that the only way to reach that almost $500 billion Black consumer market was with Black agents, like Naval Academy graduate Mike Hill of Fort Walton Beach, Florida. Allstate bought into cultural diversity because there was an economic payoff. And it paid off.

The insurance company began recruiting at Black colleges. Its managers were motivated to hire non-Whites. One-third of its promotions and a percentage of its entry-level hires were voluntarily set aside for non-Whites and women. The number of Black agents was doubled.

Allstate is now the top insurer in Chicago's Black community and in New York City. "Diversity is a business issue, not a social issue at Allstate,"[9] said Ron McNeil, a Black man who sits on the company's board of directors. As more employers face the same situation that Allstate faced in searching for new markets, they will be forced to reach the same conclusion—that cultural diversity is good for the bottom line.

Playing by the Same Rules

Some mistakenly believe that to practice cultural diversity, employers must evaluate the work performance of various groups differently. But that is not cultural diversity any more than rewarding sexism or racism is cultural diversity. Both "norming" and discrimination have adverse impacts on corporate performance. Once everyone recognizes diversity in the corporate culture, the playing field becomes level for all employees. Everyone is judged on the same standard. Everyone knows where the team is headed, and what route to take. And reasonably intelligent players should know how to help everyone on the team reach the common goal.

The concept of racial integration has never had a chance in this country because, in practice, it has always been corrupted into racial assimilation. Integration was pitched as "equal rights" but too often it has been practiced as equal rights for those who most act like Whites. This has largely undermined affirmative action programs because the Blacks who are recruited for them often are those who are most adept at mimicking European traits. Acting White is not the same as proving competence. Assimilation is particularly damaging to productivity, because it feeds cultural denial and shifts the focus from meritocracy.

Instead of being judged on performance, Blacks are too often valued according to their ability to make Whites comfortable in their presence. Once, after I had established a rapport with a White acquaintance, he remarked, "Tony, when I look at you, I don't see a Black person." I replied: "Then let me loan you my glasses." If he has to pretend that I am not Black, he has a problem with me being Black, and sooner or later his delusion is going to cause trouble for both of us. A quick reality check eliminated that prospect.

Whites may not feel comfortable being upfront about racial differences, but they can see the inefficiency in the time and energy spent pretending to ignore such differences. Often, Blacks in predominantly White settings are rewarded for cultural denial or marginalized further if they associate with other Blacks or express appropriate concern for the

219

Black community. A White woman once told me that she watched me regularly on television and felt that I was very intelligent, but, she said, "You have a Southern accent."

"Yes, ma'am," I affected. "If you'd been born and raised in Charleston, West Virginia, you'd have a Southern accent."

She apparently felt there was something wrong with the way I spoke, which to me meant that she felt there was something wrong with me. I'm sure she meant well, but patronizing relationships rather than the acknowledgment of unique differences as assets would be counterproductive if she were managing a culturally diverse group.

On the other hand, there are those who express their faith in Black inferiority as racial benevolence, such as a law firm that was described in *The New Republic*.[10] In the firm's politically overcorrect climate of liberal racism, Whites could not discuss sports with Blacks or ask what law schools they had attended under penalty of being fired. I fail to see how these draconian rules increase productivity or cause any other benefit. How demeaning it must be for Blacks to work in that environment. And rest assured that when the quality of the work falls, as it surely will, the Blacks will be blamed by the very management that has forbidden coworkers to deal with each other as equals.

The reality is that an increasingly competitive marketplace demands an egalitarian management approach. No one can work in an environment in which Whites and Blacks edge around each other like sidewalk hazards. Ultimately, I predict, the ridiculously overcautious law firm will lose market share. There is no excuse for hiring overly sensitive or incompetent Black lawyers when there are hordes of qualified ones who can give or take with the best of them. Diversity should be embraced, but incompetence should be managed as firmly as ever before. If a Black or a White is incompetent, he or she does not belong on the team. Creating a safe zone for any group of incompetents is poor management. This creates resentment, not respect, whether the employees are allowed to articulate it or not. Rage will ultimately find expression.

Democratic Capitalism

There are consultants who believe that diversity management is a nonquantifiable process. I believe, quite to the contrary, that diversity management is precisely about scoring points in some quantifiable manner. I suggest performance be measured against corporate objectives and culture. If you cannot score, you cannot play. Is our competition in a competitive global economy playing with third-string players? No. We won't either if we intend to win.

The assumption that Blacks are not qualified is racist, as is any preference program that allows White men to assume an air of superiority based on skin tone rather than performance. Because cultural diversity is performance-oriented, it gives everyone an equal opportunity. Cultural diversity is quintessential democratic capitalism. If White male managers are required to reach "diversity goals" and reap compensations when they do, and if they are penalized with a financial loss when they do not, then the entire process must reach corporate objectives or be declared a business liability. That means that women and non-Whites are equally responsible for reaching corporate objectives and they should be equally rewarded when they do.

It is true that diversity management has been skewed

by irrational demands from liberal racists and those Blacks who prefer racial assimilation to racial equality. Tired of assimilation at all costs, many Blacks and Whites have turned their backs on anything resembling integration. This is done out of resentment against the coercive policies of the past rather than out of racial animosity. And this resistance, if unchecked, will kill the process of cultural diversity, which must be viewed and understood as being in everyone's best interests if it is to succeed.

Colorizing Business

Some businesses have already grasped the fact that diversity in the workplace is not some altruistic goal. It is a matter of being competitive. The *New York Times* reported on a Maryland biotechnology firm that turned down $25 million in incentives to move to Des Moines, Iowa, because Des Moines is simply "too white."[11]

"It's becoming obvious that we have to be concerned about diversity, not just for social and moral reasons, but for economic reasons,"[12] admitted Michael Reagan, the president of the Greater Des Moines Chamber of Commerce in the *Times* story. Iowa is 96.6 percent White.

Corporate America is awakening to the fact that being all White or otherwise homogeneous is no longer an asset. It is beginning to understand that the ability to compete will depend largely on the ability to increase long-term productivity of the workforce, especially among women and non-Whites. Development of human capital leads to increased skills and knowledge.

Three Forms of Wealth

Dr. Gary Becker, a Nobel Prize winner in economics, reports in his book *Human Capital* that private rates of return on education exceed those on business capital.[13] This means that training and education are the best investments a society can make. And since the new work force will be significantly female and non-White, those groups can be ignored only to our own collective detriment. Make no mistake, neither sainthood nor philanthropy is the goal here. The bottom line is wealth creation . . . democratic capitalism.

Many of us believe that financial capital is the only form of wealth, but money is only one form of wealth. A second form of wealth is social capital. I call it your "sophistication level." For example, when going to a job interview, you put on your best suit, clean your nails, and arrive fifteen minutes early for the interview. If you want to get along in the business world, you learn how to present yourself as a part of that world. And you behave in a manner acceptable in the culture of that workplace—not in the culture of your home or neighborhood: "Yo bro!" is not a proper salutation at the headquarters of PepsiCo. Social capital is a form of wealth because it has real value in the marketplace.

The third and the most important form of wealth is your "human capital." Your human capital is the sum of your formal education and your acquired skills. Those with high levels of human and social capital are most likely to amass high levels of financial capital. Having said that, one might ask: How do so many people remain impoverished?

The answer is quite simple. Many are trapped because society marginalizes them through sexism or racism. This deprives them of the two essential forms of wealth needed to create financial wealth. Therefore, poverty is the indirect result of a paucity of human and social capital, accompanied in many cases by low self-esteem.

If you do not change the people, you cannot change their condition. And you change people with training and education, and by instilling them with moral virtues that stress personal responsibility. Strong moral character cannot be taught in a classroom, it must be integrated into the personality through personal experience. This was one of Aristotle's primary teachings. Many people lack crucial moral virtues of self-discipline, honesty, reliability, and responsibility. These are learned character traits picked up from role models. Without role models, many people never learn those moral virtues.

Centuries of racist treatment have helped impose a permanent second-class status on much of the Black community. Its roots are based in slavery and it has been institutionalized by legislation and laws. But in a modern, culturally diverse society, everyone is affected by the successes or failures of each subgroup. You do not "win" if another group "loses." That's why there are no victors in a racist society. And because there are no victors in a racist society, all of us are potential victims.

Economic Racism

What is wrong with Black and White Americans? Why can't they play for the same team? The economic and social marginality of Blacks is used as the reason for America's informal apartheid. But there has to be a reason that Whites developed a racist system that marginalizes Blacks. White racism has two dimensions: a psychological component and an anticapitalist economic one. The latter makes the concept of racism appear to be rationally derived rather than a deep-seated psychological disturbance. It legitimizes the problem, making it socially acceptable to be "normally" intolerant. And yet, it is apparent even amid the hostilities that there is a mutual fascination between Blacks and Whites.

Journalist Ellis Cose got beaten up pretty badly by some critics for saying in *The Rage of a Privileged Class* that most Whites are not racists. I agree with Cose.[14] But it is necessary to qualify that statement. I do not believe that most Whites adhere to White supremacy or Black inferiority, but most Whites do believe that any association with Blacks will reduce the quality of their lives. Even when the Blacks moving in next door earn just as much money and have just as much education, Whites move out. I believe Whites fear the market for their own homes may shrink, or that economically unqualified Blacks will follow, since residential opportunities are limited for Blacks. Some racial animosity may be in the equation, but the primary motive is economic standing.

If most Whites are racists, they are really economic racists, according to my theory. Whether racism is economic or psychologically determined, it is a neurosis, and neurotics have a greater stake in maintaining a problem than resolving it. Economic racists are not necessarily clinically disturbed; they are more or less racial and economic opportunists.

Andrew Hacker, author of *Two Nations*, informally surveyed his students and found that they believe being White is worth $1 million a year. According to this informal survey, the average White person at birth has a $70 million incentive to perpetuate racism. "And this really shows that White people do know the advantage of a White skin. And we know the experience we put Black people through,"[15] Hacker told me. This payoff for Whiteness was obvious in South Africa before apartheid was successfully challenged. Many White liberal economic racists took advantage of a racist government takeover of Black-owned farmland while professing the ideals of racial equality.

The psychological racist is not driven by material gain. Rather, the psychological racist must have someone to hate and feel superior to in order to feel personally valid. Theoretically, deep-seated feelings of inadequacy, sexual or otherwise, contribute to low self-esteem. The only relief for these inferior feelings is to maintain the delusion of superiority over an entire group of people. They must dislike an entire group of people to feel good about themselves. Examples of this pathology are seen in the behavior of skinheads, Nazis, Ku Klux Klaners, and Aryan White supremacists, among other more covert racists and, of course, millions of closet bigots.

In 1957, when overt White racism was legally reserved for the South and Black racism was in its birth stages, Norman Mailer wrote a piece called "The White Negro" in which he posed the question: "Can't we have some honesty about what is going on now in the South? Everybody who knows the South knows that the white man fears the sexual potency of the Negro. And in turn the Negro has been storing his hatred and yet growing stronger."[16]

Before I read Mailer's piece in 1994, I had already arrived at the conclusion that the fear of retaliation drove White fear of Black people and created the perceived need on the part of some Whites to keep Blacks in a subordinate, therefore nonretaliatory, position. The purpose of keeping Blacks uneducated and therefore marginal is to keep "balance," according to Mailer, who says that Whites feel sexually inferior to Blacks and so need to maintain social and economic supremacy. To give Blacks economic equality, he offers, would be to give Blacks complete victory in the war of the races.[17]

If true, that is an unacceptable trade-off in a nation fighting for its economic survival. It is a new day. When there was a surplus of superior American products fifty years ago, the average White person could afford to be an economic racist and still have a high standard of living. Today, with our fragile economy and encroaching global activity, to restrict one American from realizing his or her potential is to hold the entire country down.

Disarming Racism

My recommendation for reducing the magnitude and impact of racism is to attack it by reducing the need to see Blacks as a threat to White survival. I recommend that we focus on improving national competitiveness by enlarging the concept of affirmative action to include all Americans in need of education and training, particularly in the area of small business development. I call this an Affirmative Opportunity Program. It would eliminate federal entitlements to the middle class and level the playing field for those who need to be trained or rehabilitated for the workforce—with no regard to race or gender.

The recruitment of Blacks to lead an assault on their own social and economic problems could result in a win-win situation for the national economy, the Black underclass, and the psychological racists, who might find healthier pursuits once they understood that Blacks are not a threat. For those who depend on racism economically, an assault by Blacks on their own problems would offer everything from an expanded economy, a higher standard of living, reduced taxes, and long-term growth to, of course, less racial polarization. It would no longer be necessary to employ the anticapitalist tactic of racism for economic gain.

Global Competition

Since there is no cure for the neurosis or personality disorder called racism, in the short term our best weapon for reducing racial stress is a shared national purpose and a viable market economy—both of which are necessary if America is to remain competitive in the global economy. Corporate America is light-years ahead of the rest of the population in recognizing this fact. Many large businesses have plans for the year 2020 that include a workforce that is 50 percent non-White and offshore (meaning people who are not citizens of the United States).

Another factor is America's declining education system. If you cannot make a car in America as well as they make a car in Japan, your standard of living in America cannot be as high as it is in Japan. And you can make a car as well in America as they make it in Japan only if your workers are as well trained and educated as the Japanese. For example, many American high school students receive only one year of math. Our German and Japanese competitors receive four years of math. Many American students are lucky if they get four years of high school. If they do, they spend only half as much time on core courses as German and Japanese students.

Consider the number of days Japanese attend school each year. The Japanese go to school thirty-three days longer than Americans each year. That means in twelve years they have gone to school a year longer than the average American. Ultimately these structural problems will cost the United States its competitive edge. This decline will negatively impact every group in the country—Black, White, Brown, Red or Yellow. It has everything to do with being an American, and the solution to this predicament has everything to do with working together as a team. *Team America.*

I don't care if you don't want to marry me. I don't care if you don't want to live next to me. I don't even care if you don't want to sit next to me in school. But if you and I cannot work together and make widgets better than the workers in another country, neither you nor I will have a standard of living as high as the people in that country. We will not be able to send our children to college. We will not be able to afford a house or to get married. Under the Team America concept, you don't have to like me and I don't have to like you, but you and I have to get one thing straight: We're all we've got.

The public policies generated by this emphasis on the management of human capital will also translate politically, because they are based on an egalitarian affirmative opportunity. Everyone gets an equal chance. It is the fulfillment of

Martin Luther King's dream and the opportunity that Booker
T. Washington promised. Viewing cultural diversity as a tool
for improving the quality of the workforce is fundamentally
an economic issue—the development of human capital and
the management of human resources.

This is the essential difference between cultural diver-
sity and preference programs such as affirmative action for
women and non-Whites. Preference programs automatically
phase themselves out, because demand for high-tech skills
drives workforce decisions. Therefore, equal opportunity, not
government fiat, will drive equal outcome for members of
all groups, which will result in productivity growth and an
improved racial climate.

That is why I say that cultural diversity is America's
industrial salvation and a harbinger of the end of institu-
tional racism and sexism. Personal prejudices will linger.
There will be some last-ditch efforts to preserve sexism and
racism and Black dependency. But in the end, the need for
diversity will overcome and eliminate those long-lived social
burdens. As I like to say, now a woman's place is in the
House. And the Senate.

White males, who also constitute a diversified group,
will still be valued in a culturally diversified workforce, but
they must meet the ever-rising skill requirements in the
leading-edge occupations—the identical criterion for all
workers—and they must be managed as a diversified group:
young and old, educated and uneducated, responsible and
irresponsible.

In fact, cultural diversity will render gender, racial, and
ethnic differences inconsequential, because non-Whites and
women who are educated and trained for the jobs of the
emerging knowledge sector will be actively sought after for

high-paying positions. "Empowerment is more than a trendy slogan. It's competitive advantage," says J. Roger King, PepsiCo's senior vice president of personnel and a strong proponent of a culturally diverse workplace.

The most viable concept of cultural diversity is one that allows people of all kinds to reach their full potential in the workplace while in pursuit of corporate objectives. The corporate bottom line is cultural diversity's primary objective, not historical payoffs through preference programs or multiculturalism. But to succeed, business must manage diversity by first recognizing it.

My Team America proposal suggests an additional objective—to increase the nation's industrial competitiveness. We can remain racially separated and maintain a healthy, productive nation, but we cannot continue to disrespect or disregard our differences if we are to be competitive in the world community. Social engineering will drive itself; a healthy economy lights the way.

If you decide to live, marry, or worship outside your particular ethnic, racial, or religious group, we can talk about it. Those areas must, however, remain a matter of personal choice. And they are certainly off-limits to the government. Once we become relatively equal economically and educationally, the transition to social intercourse will be a natural process. Otherwise, and until then, no matter how much "getting to know you" we do, inequality will prohibit any meaningful relationship and can only lead to more paternalism—which, in turn, inspires both Black and White racism and ethnocentrism.

Each group must earn the respect necessary to make equality possible. Team America should provide a moral foundation to bolster our national character. It would be enlightened if we decide to help our troubled and depressed communities revive themselves. The key is helping people help themselves—the moral virtue of self-help, not liberal racism or government paternalism—and certainly not conservative meanness. That is where the American character kicks in as our most valuable asset.

CHAPTER 10

POLITICAL DYNAMITE

*The Negro has nothing but "friends" and may God deliver him
from most of them, for they are like to lynch his soul.*
W.E.B. DU BOIS

Ron Brown, Secretary of Commerce during the first Clinton administration, was a millionaire Black Brahmin. As the former head of the national Democratic Party and one of the ultimate political insiders, he was a bona fide member of the highest caste of Black America's Talented Tenth and its political oligarchy.

And typical of most of America's Black leaders, Brown had no tolerance for any Black who dared to act independently or otherwise wandered from the liberal plantation of the Democratic Party. Here was Ronald Brown's standing order on how to deal with Blacks who did not follow or vote along

Democratic Party lines: *We need to get up in their faces and embarrass them. . . . Call them at home and let them know they can never forget where they came from.*[1]

Like any Black who dares break ranks with the totalitarian Black leadership in this country, I have felt the heat—not from Brown, but from other equally intolerant Black leaders. And I say that if it takes an internal conflagration to make myopic Black Americans see the political light, well: Burn, baby, burn.

I was the subject of an exploratory committee for the 1996 Republican nomination for President. That really raised the Black brownshirts' ire. Leaving their well-guarded ideological plantation in the first place was bad enough.

It was on July 25, 1991, at a Manhattan reception held on my behalf by Lionel Hampton, that I made the announcement that got the fires burning. Except for a brief fling with the Democratic Party in my youth, I have never been a fall-in-line Black, politically or otherwise. For most of my adulthood, I have maintained an independent political status.

That independence had not endeared me to those Black elitists who have anointed themselves as leaders, the BUM. It really incited their wrath when I announced that I had decided to go a step further from the plantation and register as a Republican.

Yes, I am a Black Republican, which is an oxymoron in the minds of many Blacks and Whites. My announcement was such a departure from tradition that it made headlines. Under one that read, "Black Spokesman Wrong to Join GOP," Gannett News Service columnist DeWayne Wickham, a Black who is obsessively anti-Republican, wrote that I had "crossed the Rubicon of American politics."[2]

Ten years ago, I would never have dreamed of becoming a Republican myself. But I did not join the Republican Party to align myself with the ultra-conservatives and racists who hijacked the Republicans' image at the 1992 Republican national convention in Houston. They are aberrations, not representatives of the Republican Party that I know, or the Republican Party that I envision.

I joined the Republican Party because I know its history is rooted in the highly moral principles of Abraham Lincoln and Frederick Douglass. They and Republicans like Jack Kemp are the Republicans whom I proudly align myself with. They are Republicans that any Black should gladly stand alongside. But many Black leaders have forgotten the values that these Republicans stand for.

My concerns about the failure of America's Black leadership are no different from those of many Blacks, most of whom prefer to keep their feelings private because they are not "politically correct" among the Black elite. It is out of ignorance of their own history that many Blacks demean the Republican philosophy and condemn Black Republicans.

Blacks have been Republicans historically. Frederick Douglass and the first twelve Blacks to serve as U.S. Congressmen were Republicans. And Congressional White Republicans were the architects of Reconstruction, a ten-year period of unprecedented political power for Black people. Democrats working hand in hand with the Ku Klux Klan gave us Jim Crow laws that effectively reenslaved Blacks.

Early in this nation's history, the GOP championed legislation that removed the shackles from our Black ancestors. Republicans spearheaded a movement to force Whites to give equal rights to the former slaves and initiated the Thirteenth Amendment, which outlawed slavery; the Fourteenth

Amendment, which guaranteed Blacks citizenship; the Fifteenth Amendment, which extended the right to vote to former slaves; and the first Civil Rights Act of 1866.

If you know this history, you have to wonder: How did Blacks move from the party that gave them civil and political rights to a previously all-White Democratic Party? How did they come to join forces with a party with a history of racist demagoguery, support for slavery, Jim Crow, and lynching?

The Black movement away from the Republican Party began during the Depression, when the social programs of Franklin Roosevelt attracted Blacks to the Democratic Party. Although they were quiet about it for the most part, many Blacks remained loyal to the party of Lincoln over the years; 40 percent of Black voters cast ballots for Dwight Eisenhower in 1956, and 32 percent voted for Richard Nixon in 1960.

Members of the Black middle class who continued to embrace the concept of self-sufficiency stayed with the Republican Party more than poor Blacks did. In some prosperous areas, Republicans were getting nearly 50 percent of the Black vote as late as 1960. Then Democrat Lyndon Johnson won over many Black voters as he enacted his historic civil rights legislation and accomplished more for Blacks than any other President in American history. With Blacks supporting Johnson, disgruntled Southern Whites defected to the Republicans in the 1964 election and the Republican version of the Southern strategy was born.

Breaking with Party Lines

When I crossed over to the Republican Party, I was gambling that I could help the party return to its roots and find its soul. I was quickly made aware that most people do not approach politics in such a strategic manner. Many Blacks have questioned or openly condemned my decision without really contemplating my motives. Others were simply perplexed. "Brother, you used to be my main man," a Black caller to Chicago's WVON-AM radio said. "You're still my man, but why'd you have to become a Republican?" The reception was not a lot warmer from the other side. Pluria Marshall, a Black and a longtime Republican, had a typical reaction. In a *USA Today* article entitled "Noted Commentator Joins Republican Ranks," Marshall wondered if it was "an opportunity for the Republican Party, or an opportunity for Tony Brown."[3]

At the least, my alignment with the Republicans produced some rare racial unity—liberal Whites attacked me too. The White extreme left hired an anti-Black Black, writer Playthell Benjamin, to vilify me personally in the *Village Voice*. And in another interview a White reporter bluntly asked me: "Why would you as a person who has spent his life fighting racism join a party of racists?"

Dole and Gingrich Welcome Diversity

That biased and uninformed view of the GOP was proved wrong just a few months after my announcement, when the then Senate minority leader, Bob Dole, hosted a breakfast in my honor with the Republican leaders of the Senate. Dole, who later became Senate majority leader, was primarily interested in developing a new and legitimate legislative approach to bring Blacks into the mainstream. Shortly after that breakfast, a little known Congressman from Georgia invited about twenty-five Republican leaders in the House to meet with me at a luncheon.

Newt Gingrich was then the minority whip of the House. He was curious about me and my motives, so we spent four hours together that day. In many visits since, I have found Gingrich to be a visionary of the third wave who believes that "American civilization" will be destroyed unless we give priority to the inner cities. I also discovered that both of us share an academic background and an iconoclastic approach to public policy issues.

Gingrich is not a bad fortune teller either. On our first visit, he took me into the office of the Speaker of the House, then a Democrat, and informed me that he would one day hold that title, occupy that office, and enjoy both the grand view and the power it offered. "It is the battle of ideas that is at the heart of our political order," he told me.

When Gingrich introduced me before I addressed a national GOPAC meeting, he told the audience: "Tony Brown is a genuine historic figure—and he is going to make history." He suggested that my motive in joining the Republican Party is to keep both parties honest. In fact, he said, Republicans should be aware that "we can count on Tony Brown to let us know when he thinks we are wrong."[4]

Gingrich asked me to speak at Republican gatherings around the country. On one occasion, when he was winning reelection by a very narrow margin, he asked me to speak at an Atlanta luncheon for his major financial contributors, all of whom were White. On this occasion, as on others, he never asked ahead of time what I intended to speak about. In fact, I was pleasantly surprised at his adherence to the First Amendment and my challenge to make the GOP more inclusive, balance the budget, drop the Southern strategy for Team America, and get back to the party's historic roots.

Interestingly, the general public seems to sense the truth about my political affiliation too. The response from the public was overwhelmingly supportive. A prominent Black woman in Baltimore, the wife of a minister, followed my lead and registered as a Republican. The *Wall Street Journal* invited me to write an editorial explaining my decision, and when I gladly complied, they gave it prominent play. In the editorial, I noted that after my announcement, a surprising number of Blacks told me they planned to follow my example.[5] And all of the Blacks who have spoken to me, whether they agree or disagree with my decision, admit that they like the idea of greater ideological diversity within the Black community. That sentiment shows up in one opinion poll after another.

The statement that best typified the reaction to my becoming a Republican came from a woman at the Apollo Theater in Harlem, following my interview on WLIB, a New York Black radio station where I now host a daily talk show. "When I first heard the news, I thought you had sold out. So I had to hear an explanation from you personally. After hearing you explain, I agree with you and admire your courage. We do have to rely on self-help and we do have to be involved in both parties. But I'm not ready to become a Republican yet. I don't trust the Republican Party, but I do trust Tony Brown."

The Truth Wins a Round

In one of the more interesting invitations to come from my controversial move, the fiery activist Al Sharpton asked me to join him in a debate on political choices for Blacks. A huge audience packed a school auditorium in Harlem on a hot Saturday morning in August in anticipation of a political slugfest between the Reverend Al and myself. I spoke first. With tongue in cheek, I congratulated Blacks on the large salaries, wonderful new homes, safe streets, excellent schools, and plentiful jobs they had created for themselves in Harlem over the years by being such a solid voting bloc for the Democratic Party. "Aren't you grateful for all of the welfare people say that you're getting?"

They got the point. And the potential Brown-versus-Sharpton slugfest turned into a lovefest as an amused audience roared with laughter, approval, and self-recognition. They came to realize, as I had, the hypocrisy and reality of the Democratic Party, or, for that matter, depending exclusively on any political party.

Sharpton then agreed that Blacks need all the help they can get from anyone willing to give it. He accepted my point that if Blacks are not prominent in the Republican Party, then they cannot change it so that all people benefit. In conclusion, Sharpton said that he would continue to work for the benefit of Black Americans in the Democratic Party. I plan to do the same in the Republican Party. End of the debate. At least that one.

One prominent Black Republican, Gary James, president of the African-American Republican Task Force, later raised a point that Sharpton and others among my critics had missed. James asked in a letter, "Tony, why are you going to all of this trouble to recruit Blacks for the Republican Party when there is no payback? The White Republicans will simply do what the White Democrats are currently doing—take Black voters for granted—if we help them win back the

Congress and the White House. It escapes me how Blacks would be empowered . . . in the unlikely event that the GOP took your advice and regained the White House in 1996. The power dynamic for control of the Republican Party will continue to be the essential preoccupation of party politics, your laudable lobby notwithstanding."

I assured James that I had not overlooked this "hole" in the potential plot development. I have developed a strategy beneficial to both the Republican Party and the Black community.

The Tony Brown Solution

As corny as it may sound, I believe that the ultimate power is in service to others. My win-win strategy is based on the fact that if the Republican Party wants more widespread Black support, it must become an inclusive party and the party of affirmative opportunity. Concessions can and must be made without surrendering the Republicans' basic principles of limited government and free-market solutions. The Black community must assume a leadership role in developing those solutions.

If Blacks want power within the Republican Party, Blacks themselves must initiate change. If both Republicans and Blacks are willing to accept new roles, they will both succeed politically. And if those changes are made within the Republican Party, the Democratic Party will be forced to accommodate broader Black interests, because for the first time in decades Democrats will have to compete for the Black vote rather than taking it for granted. If the Republican Party demonstrates that it embraces affirmative opportunity,

Black voters who are looking for private-sector and entrepreneurial solutions will have an alternative to the tax-and-spend government dependency offered by the Democrats.

The sleeping Black giant of American politics has been conned by the BUM into blindly voting for any Democrat without receiving anything in return. And you can't sell what you give away. By increasing its options with representation in the Republican Party, the Black community will free itself of the BUM's hammerlock. The key to this scenario is a new flexibility. Blacks must be willing to exercise their new options.

As the quintessential Republican Frederick Douglass said: "Power concedes nothing without a demand." Blacks cannot keep on doing what they've been doing and expect to get any better than what they have been getting. We must change our tactics in order to bring change in our lives.

The Republican Party cannot give Blacks political power. We must earn it through our own initiative and creativity. Political empowerment does not come if your party wins without your support. You have to be a player. Your support, broadly defined as the number of votes you can turn out, is your only bargaining chip—but to have maximum impact, it must be managed on behalf of your supporters. Jesse Jackson's candidacies in 1984 and 1988 failed to do so.

On the Presidential Ballot

As I said, I am the subject of an exploratory committee for the 1996 nomination of the Republican Party for President. If I do run in the primaries, I will call for renewal of the American character, shared sacrifice, and fairness for all.

With Jack Kemp's decision not to run for the presidential nomination in 1996, there is a vacuum for these ideas among Republican aspirants and in the party.

Many Republicans might be energized by such a candidacy, and many disheartened Whites who had once been Republicans might return to a party that had rediscovered its soul. I believe that it will surprise Americans when they learn how free of psychological racism most White Republicans are and how they too might welcome the opportunity to demonstrate their fairness and tolerance while embracing deeply held principles.

Some political observers employ a double standard in finding Democratic racism less offensive than Republican racism. But I agree with those who call for Blacks to position themselves so that both parties have to court the Black vote. I'm all for that, but it certainly cannot happen unless a large number of Blacks become active in the Republican Party, or in an independent political movement.

Already, there are trends that might ignite political dynamite and blow apart the Democrat plantation where Blacks have been taken for granted. While there is a growing hard core of Black anti-Democratic sentiment—about 20 percent of the total Black vote—the Black vote has become of increasing importance. Consider that since 1968 the Democratic presidential candidate has not carried the White vote. In 1992, Republican George Bush won a plurality of the White vote, but lost to Democrat Bill Clinton, who received a mere 39 percent of the White vote. And yet the presidential election of 1992 was the first since 1944 in which the civil rights of Blacks were ignored as a platform issue—by both parties.

During the 1992 presidential campaign, the Democratic

Party, under Ron Brown's leadership, resurrected the "Southern strategy" that it originated during slavery. Under this divisive strategy, Clinton made no direct appeal to the Black voters, even though he was relying on them to carry him into office. The idea, according to Professor Charles V. Hamilton of Columbia University, was to conduct a "visible contest for the votes of whites and an invisible quest for the votes of blacks"[6] because White Democrats resented direct appeals to Blacks.

Hamilton noted that this strategy was formulated and accepted by Brown and many other Black leaders. The BUM may never have that opportunity again. Blacks have quietly been either staying home on election day or voting in large numbers for White and Black Republicans. Blacks silently fleeing the Democratic Party contributed mightily to the Republican blowout in 1994 that allowed the GOP to take control of Congress for the first time in forty years.

Representative Harold Ford from the mostly Black ninth Congressional district in Memphis threatened to retaliate against anyone who voted for his Republican opponent, Rod DeBerry, a Black who pulled 42 percent of the vote in 1994 in what amounted to Ford's most serious challenge in eleven terms. Ford threatened to cut off services to the Republican-voting dissidents. Enraged, he said, "I know where the devils live."[7] Black Republican Marc Little of Jacksonville carried 42 percent of the vote also in his Florida race, and barely missed a huge upset in a majority Black third Congressional district. Ron Freeman, another Black Republican and near winner, carried 43 percent of the votes in the Kansas City, Missouri, fifth district.

The number of Blacks who want an independent Black political party has doubled in the last five years, according to surveys.[8] These developments are the makings of political dynamite that could open the way for a political revolution. Moreover, if you can put the pieces together correctly, you'll get a whole that is much larger than its parts.

Throwing the BUM Out

Black leadership in this country has scrupulously avoided having a plan of action, a road map to equality—either because they are not smart enough (which I doubt is the case) or because they don't want Blacks to leave the Democratic plantation. Ron Brown's "in-their-face" directive cited at the beginning of this chapter is very explicit on the latter point.

On June 18, 1971, writer, author, and actor Ossie Davis was eloquent as the keynote speaker at the Democratic Congressional Black Caucus dinner. He told the audience of almost three thousand of the Talented Tenth—among the most educated and gifted of Black professionals—that his message was "very simply: It's not the man, it's the plan." Davis noted also that rhetoric will not solve our problems with this succinct summation: "It's not the rap, it's the map."[9]

Davis looked straight at the Black members of Congress and challenged them: "We need a plan. And that's why tonight the burden of my appeal to you—the thirteen Congressional Black Caucus members—is to give us a plan of action. Give to us a Ten Black Commandments. Simple. Strong. Something that we can carry in our hearts, and in our memories no matter where we are . . . a simple, moral, intelligent plan that must be fulfilled in the course of time; even if all of our leaders, one by one, fall in the battle, somebody will rise and say, 'Brother, our leader died while we were on page three of the plan, now that the funeral is over, let us proceed to page four.' "[10]

To this day, the BUM's elitist members have never turned the page. They have never come up with a plan of action. As a result, almost thirty years later, Black people are stuck on page one, which is entitled "Modern Slavery." We are slaves chained to one party in a two-party system. We live on welfare plantations. We are burdened with one-way integration and noneconomic socialism. And this is all the result of the policies of the self-serving elitist BUM.

Leveraging Black Political Power

If the Black Unaccountable Machine were truly committed to the political empowerment of the Black rank and file and not devoted to obliging the White left, then its self-promoting members would demand compensation from the Democratic Party for the *strategic* voting power of the Black bloc instead of giving it away.

In fact, the strategic distribution of Black voters in the dominant nine electoral college states has previously determined winners in the national race for President. The Black vote in key electoral states canceled the national majority White vote against Jimmy Carter in 1976 and put him in the White House. In 1992, Democrat Clinton again lost the White majority vote, but won the Presidency because Blacks provided the winning margin with 82 percent of their strategically placed votes.

The majority of Whites abandoned the Democrats long before the blitz of 1994. In the end, 63 percent of White men and 59 percent of White women voted for either George Bush or Ross Perot.

In the same national election, George Bush failed because he ignored a strategy that would attract the many Blacks who believe in self-help, a market economy, entrepreneurial opportunities, limited government, personal responsibility, a balanced budget, and an inclusive society. If Bush had adopted that agenda, he would have garnered at least 20 percent of the Black vote. To show how strong the basic message of Republicanism is to a traditional Black culture, Bush carried 11 percent of the Black vote against Clinton with no effort. Another 9 percent of the Black vote would have kept the Republicans in the White House.

Here is a plan to get Blacks to demand concessions from both major parties in exchange for their votes. In presidential elections, a candidate needs 270 electoral votes to win. Nine states alone control 243, or 90 percent, of the necessary 270 electoral votes. These Big Nine states are California (54), Florida (25), Illinois (22), Michigan (18), New Jersey (15), New York (33), Ohio (21), Pennsylvania (23), and Texas (32).

As you can see, three electoral states in the North dominate the electoral system: New York, Ohio, and Pennsylvania. These three states alone provide almost one-third of the electoral votes to win the national contest for President and all have substantial Black populations, as do most of the Big Nine. A shift of a swing-voting bloc into one candidate's column and out of another's in nine states alone can decide which party controls the White House. If the Black vote would act as a swing vote, then Blacks would exploit the natural division in a two-party system.

A third-party candidate would enhance the strength of a swing-voting bloc even more. If, for example, Ross Perot had decided to organize his United We Stand Party as a swing vote in 1992, he could have handpicked the next President. In that unlikely event, it would have made it even more imperative that the Republicans attract a larger Black vote through primary participation—now around 1 percent of the total—and by putting a Black at or near the top of the Republican Party ticket.

Of course, the Republicans won a majority in 1994 without a substantial Black vote, adding to their political irrelevance. While a moderate Black such as Colin Powell may beat President Clinton in the opinion polls among Whites, most Whites are fearful of Black Democrats in Congress because of their liberal voting records.

Roll Call ranked twenty-four of the Black Caucus members (three-fifths of the entire group) as "ultraliberal." It ranked 85 percent as "certifiably liberal." The Democratic Congressional Black Caucus, which is essentially a socialistic body, is generally identified in the media as the "liberal wing" of the Democratic Party. Blacks are only 9 percent of the House, but they constitute 50 percent of that body's extreme left—the top 10 percent.[11] Even when Congress is focused on debt reduction, they are the biggest of the big spenders. This is an untenable position.

The portrait of Black liberal leadership is framed with irresponsibility. Therefore, considerable doubt exists about the ability of Blacks to hold and exercise power. Even a moderate Ronald Brown type (without his ethical and legal problems) on the Democratic ticket would likely hurt the presidential candidate because of the White fear of the Black Democrats in Congress. Two-thirds of Blacks, according to a University of Chicago study, identify themselves as either moderates or conservatives, and many of them must privately share concerns about the reckless spending by most Black Democrats in Congress.[12]

With their noneconomic socialist philosophy and their reckless attitude toward the financial and fiscal management of the government, it is abundantly clear that the Democratic Party and the Democratic Congressional Black Caucus cannot effectively represent the interests of most Black people. Under the present monopoly-plantation system, Blacks in Congress can get away with voting for legislation that many Blacks oppose—tax increases, soft crime programs, welfare programs that encourage teen girls to have babies, or larger deficits—as long as they can convince Black voters that the devil is a Republican. Believe it. During the Republican sweep in 1994, not one Black Congressional incumbent lost.

The BUM controls the Black vote, and the BUM itself is controlled by the White left. I doubt that outsider Louis Farrakhan understands that much of the enmity toward him in the media is not generated so much because of his anti-Jewish and anti-White sentiments—although those inflammatory stands certainly don't help. What has really earned him their wrath is his potential as a political threat to the White left's control of the national Black vote.

Farrakhan's following among the most disaffected Blacks threatens the power base of the Black elite. The votes held by poor and exploited inner-city Blacks into the hands of Farrakhan would automatically become a swing vote. Farrakhan's attempt to put one million Black men in a Washington, D.C., march is a conspicuous attempt to enshrine himself as America's premier Black leader. A fear of Farrakhan's potential political power is the force behind the "covenant" with the ten-thousand-member Nation of Islam that was announced by the previous Black Caucus leader, Kweisi Mfume.

If there is little help forthcoming for the Black community from its Congressional leadership in the arena of electoral politics, the lack is even more acute in the area of business and economic development. Normally, Black business owners would look to the Democratic Congressional Black Caucus for help to halt the Democratic Party's punitive regulatory and tax legislation, but it has become very evident that Black Caucus politics are not the politics of business or self-reliance.

The Black Caucus's most publicized emphasis on economic development is its support for "empowerment zones." This, in reality, is a plan to give mostly non-Blacks tax breaks to open up more businesses in Black communities. While that is acceptable and a few jobs are better than nothing, how that will "empower" Blacks or create economic self-sufficiency escapes me.

Endangered Black entrepreneurs who truly empower Black communities supply 80 percent of all new jobs created in their neighborhoods, but they don't fit the noneconomic socialist agenda of the Black Caucus or the Democratic Party. Nor do they fit into the current Republican agenda.

The same Republican conservatives who demand wealth creation from Whites define "the next stage of the civil rights revolution" as building strong families and communities among the Black poor and the working class. How in the world can you build strong families, communities, and reform schools without economic self-sufficiency? No other group ever has. Racism is debilitating, but poverty is the major predicament facing Blacks. Blacks desperately need economic solutions. Self-employment is a time-tested economic solution.

Economic self-sufficiency is the foundation upon which society is built. The value of a family to a society is that it makes the capital investment in the individual members so that they, in turn, become contributors to the economy and help drive national wealth. Entrepreneurship is the catalyst for self-sufficiency among those who are poorly positioned, and generally that means the impoverished and the immigrant groups.

The Jews, the Irish, the Pakistanis, the Koreans, and the Chinese have embraced entrepreneurship, whether through home-based businesses, produce stands, or pushcarts. Entry-level enterprises led to Macy's and Gimbels. The national economy is founded on entrepreneurial endeavors that create wealth in the communities, not by empowerment-zone tax breaks for those who have already secured their wealth.

And that's precisely why the GOP lost the White House in 1992. Republicans historically have not targeted Blacks who believe in self-help—conservatively estimated at 20 percent of all Blacks who vote. Indeed, White Republicans are too intimidated to even debate Black Democratic leaders; they fear being called racists.

Some Black liberals shout "Nazi!" at any Whites or "Uncle Tom puppet" at Black journalists and other Blacks who dare to challenge liberal socialism. This effectively neutralizes all criticism and places any legitimate political difference in a racial context.

Unfortunately, Republican emphasis is on placating the BUM to keep its members happy and to keep their anti-Republican attacks at low volume. This strategic acquiescence to the fear of being labeled racist and the White culture's inability to deal with Blacks as individuals are as counterproductive as the exclusionary policy of the Southern strategy.

The Politics of Soul

As a result of Democratic demagoguery and hegemony over most Blacks politically, many Republicans don't feel a dependable share of the Black vote is worth the effort. One wing of the Republican Party believes that Blacks have been so corrupted over the last thirty years by liberal pandering and "compensatory deference" that their values and absence of the moral virtue of responsibility preclude their qualification as Republicans and as true conservatives. To win them over, Republicans would have to join the pandering game and surrender their own core values for strategic political gain, the logic concludes.

If we are discussing carrying the majority of Blacks, trying to turn geese into swans, I concur. In precise philosophical terms, most Blacks are too liberal to be Republican. However, polls strongly suggest that their emerging independent political streak is hardening into a swing vote. This movement away from a near complete manipulation by the Democratic Party can be tapped into spectacularly by the GOP with an agenda that is fair, yet not race-based but culturally sensitive and relevent to the nation's growth.

Even culling the Black vote by pandering to upwardly mobile and wealthy Blacks with preference programs fragments the Black community further and exacerbates its socioeconomic predicament. Class pandering also produces more dependence and racial polarization. Conversely, there is considerable overlap between racial audiences if fairness and opportunity are political objectives.

I propose a new GOP agenda of opportunity and fairness that could transcend racial and political barriers and simultaneously attract philosophically compatible Blacks from all

252

socioeconomic groups. Republicans may be amazed at what a party with a little "soul" can accomplish politically. Most Blacks, for example, will see the wisdom of affirmative action for the socially and economically disadvantaged, the merits of repealing the Davis-Bacon Act, the benefits of a repeal of the income tax, the need for welfare reform that protects children, and they will applaud an official government apology to Americanized Africans for centuries of America's worst atrocity—slavery—and the psychological and economic legacy it imposed on the descendants of slaves. The blame for the damage to Native Americans and Japanese-Americans was accepted by the U.S. government and compensation was paid. Germany has apologized and paid five billion deutschmarks a year to the Jews for its atrocities since the end of World War II. At least the United States owes Blacks an apology.

A gesture of this type by the GOP would begin the national healing and guarantee its status as the majority party. Will the GOP read the Black community perceptively, see that times have changed, and boldly reform itself? Or will its members become cocky because of the 1994 sweep and so persist in their exclusionary dogma?

Author and conservative Kevin Phillips and conservative columnist Samuel Francis say that the Republican Party does not need the Black vote because it carried 63 percent of White male voters in the political lobotomy the voters performed on the Democrats in the November 1994 elections. By remaining "a mostly White party . . . keeping its commitment to White voters," Francis argues, and not "pandering to non-existent Black Republicans" as Newt Gingrich and Jack Kemp propose, the GOP will have a future with "its real voters."

In 1996, however, this racial exclusivity could be costly. In addition to alienating nonracist, moderate, and middle-class White voters with an intentionally exclusive policy, if a new center-right party is formed by the Perotista types who abandoned the Republicans in 1992, it would amount to a one-third loss of the White vote for the Republicans. And with the Black community energized by a GOP anti-affirmative-action campaign, Black liberals would once again convince many Blacks that Republicans are racists. With 90 percent of Blacks voting Democratic in this scenario, the Republicans could not recapture the White House and would probably lose control of Congress. Only an independent Jesse Jackson run for President that would siphon off 10 to 20 percent of the Black vote from Bill Clinton in the general election could save the White House for the Republicans in that scenario. Don't bet on it though. Jackson usually places party interests above race matters.

On the other hand, a 20 percent share of the Black vote would win the Presidency and carry a substantial number of Congressional districts for the GOP, because in ten or twelve states the Democratic candidate cannot win with a Black voter defection of that magnitude. It would also substantially increase the GOP Congressional majority and introduce a Republican Congressional Black Caucus. Blacks are poised to vote Republican in much larger numbers if the GOP is sensitive enough to harvest the opportunity.

Phillips and Francis are correct in the sense that the Black vote as a bloc is politically irrelevant and Black socialists will never make good Republicans. In fact, Blacks have become politically irrelevant because they vote as a bloc. The Republican tsunami of November 8, 1994, captured Capitol Hill without 88 percent of the votes cast by Blacks; the Democrats couldn't defeat a single incumbent Republican governor, senator, or House member anywhere in the country even with an 88 percent Black-bloc vote. Republican ideas won control of both houses of Congress, the majority of governorships, and an impressive increase in state legislatures.

Many Whites will defect from a party stigmatized as racist. After the 1992 Republican national convention in Houston and the televised spectacle of Pat Buchanan's out-of-step nativism, the party of emancipation was openly called divisive, mean-spirited, and racist. Many suburbanites, working women, gays, pro-choice advocates, and Jews, and even some moderate WASP Republicans, came to feel as unwelcome as most Blacks. Pollsters later told the nation's GOP governors that voters added "narrow-minded" and "restrictive" to the image of Republicans. Buchanan's later-stated dream of the day when "hard-eyed" men in boots will take over Congress has not attracted any tolerant voters either.

Cultural diversity (not to be confused with multiculturism or affirmative action) is the foundation of our economic competitiveness. One of the biggest hurdles to an inclusive GOP party is that many conservatives have a centrist concept of "American" conduct and tend not to accept diverse cultural nuances. They resist appealing to the uniqueness of population sub-groups. Therefore they are often perceived as racist and indifferent, which is the most consistent criticism of Republicans by Blacks. This rigidity shows up in political appeals to Blacks and results in Republican failure to break the Democratic stranglehold on the Black vote.

The Three-Party System

It is abundantly clear that the Black leadership elite is incapable of executing a viable plan to alter the miserable situation for the Black masses. That is why one possible solution is at least three vehicles of political expression: Republican, Democrat, and Independent.

Let the Independent wing, run by an independent presidential candidate like Ron Daniels or some equally competent socialist-progressive, organize around issues such as reparations for slavery, Pan-Africanism, cultural identity, and related concerns that are liberal-Black, not liberal-racist.

Fire the standing Democratic socialist Black leadership and replace it with Black Democrats who will represent the Black poor by pushing for temporary welfare benefits for the needy—along with a pragmatic liberal agenda that emphesizes a major role for government in helping poor people. And finally, demand that Black Republicans push within the GOP to move Blacks out of poverty and off welfare into opportunities that will raise them into the middle class.

If Black Americans direct their political energies toward this agenda for empowerment, political parity for Blacks will be achieved. If the Black vote were split 50 percent Democrat, 25 percent Republican, and 25 percent Independent, the Black community, which is already the majority voting bloc in many major urban areas, would be a political powerhouse.

As it stands now, however, the Black community's political power is unfocused and untapped. The BUM blames the decline of the Black community on Republican racism, indifference to poor people, and twelve years of Republican Presidents. But during those twelve years, Blacks did not vote for Republicans and the party they did vote for did not protect them, except for payoffs to the Black elite.

For forty years, Congress was controlled by Democrats. If Blacks had voted heavily for Republicans over those four decades, would Republicans not be accountable, in part, for their socioeconomic status today? Blaming the Republicans for failing to do the Democrats' job is like blaming the winning team for scoring points. Whatever the reasons, your team only loses when it scores fewer points than the winner.

Politics is a quid pro quo business—if you help me, I pay you back. You do not oppose a party and demand the spoils if they overcome your opposition and win. And even with the Democrats in control of Congress and the White House during the first two years of the Clinton administration, conditions for ordinary Black people showed less dramatic improvement, certainly no more, than during the Republican White House reigns of Reagan and Bush.

Jim Crow in the Clinton Camp

Prior to his election, crafty Clinton capitalized on the Sister Souljah incident as a way to take a not so subtle slap at Jesse Jackson and thereby send a message to ease the fears of White voters. He let them know that Blacks would be kept in line if he won election. Clinton even flew back to Arkansas to witness the execution of a brain-damaged Black man in order to portray himself as hard on crime—that is, on Black criminals.

With those acts, among others, Clinton provided the country its first look at a "New Democrat" who in reality was just a refried liberal version of George Wallace and the old White-supremacy South. Starved for power, White Democrats went for the presidential combo platter. After those exhibitions of pandering to White racism at the expense of Blacks, Clinton was never again behind in the polls.

His plantation political strategy was so adroit that he garnered 82 percent of the Black vote and his share of a three-way split of the White vote. His scam was so perfect that his victims guaranteed his victory. In 1992, George Wallace III won the White House, but with a twist of liberal racism. After the Republican landslide in 1994, Clinton promptly reconverted to a Wallace Republicrat.

Both parties recognize a fact that was demonstrated in the elections of the last four Presidents. It is that most Whites feel no further obligation to assist Blacks and are adamantly opposed to the new version of civil rights that guarantees success, not opportunity, for Blacks.

It is demagogic propaganda for Blacks to blame the Republicans when they race-bait while praising Democrats for doing the same thing. The truth is that both parties play to the prevailing mind-set of the White majority. Black failure and White racism did not begin with the Republican Party of Ronald Reagan or George Bush. Neither were the "gains" of the civil rights movement in the 1960s reversed because Republicans won the White House in 1980 and will very likely win it again in 1996.

A little recent history will be more informative than the same tired alibis that Blacks get from the BUM's leadership. The White majority controls America. Black advancements given by Whites in the sixties and seventies were reversed in the eighties and nineties by majority rule, in part because many Whites became convinced of "reverse discrimination" by Blacks, and in part because of the loss of so many middle-class jobs. A history of racism underscored those fears.

Republicans won in 1980 because the White majority,

which has not voted for a Democratic presidential candidate since Lyndon B. Johnson in 1964, wanted Black advancements reversed. Both major political parties became vehicles for that expression, and more recently the U.S. Supreme court. The fact is that the Republican Party of Reagan and Bush was empowered by the White majority to halt Black preference programs, as was Democrat Bill Clinton later on.

While the GOP exploited the anti-Black mood, as did the Democratic Party whenever it could, it did not create the tensions. It did not have the capacity to do so on its own. The tensions were aroused as a consequence of group interests and group conflict. The insistence of Black leaders on guaranteed success triggered White racial fears, which are never too far from the surface.

Middle-class and marginal ethnic Whites, who were already suffering from a loss of real purchasing power because of technological displacement and structural changes in the economy, had to hold on to what was left after Blacks began moving into the workplace under affirmative action programs. The years of fearing Blacks and racial isolation fanned the flames of suspicion and resistance and resulted in the "reverse discrimination" backlash. Whites needed a scapegoat to explain America's economic decline.

The left has a unique way of playing the race card to maintain a liberal power base. The foundation of the liberal establishment's entire power base is the fact that Blacks have had a blind faith in White liberals. Bill Clinton offers a textbook case of the exploitation of this misplaced loyalty.

In the *American Spectator* article by David Brock detailing then Governor Clinton's sexual escapades in Arkansas, a state trooper told the writer that Clinton revealed his strategy for getting elected and reelected in Arkansas. "If he could hold the black vote, generally about 18 percent in a state election, his victory would be sealed," Brock quoted the state trooper as quoting Clinton. The trooper added that Clinton told him "that meant his opponent has to get his 51 percent out of 82 percent" of Whites.[13] Shave the Democrats' share of the Black vote by another 10 percent and the liberal empire will topple in short order.

Many forget that Clinton is a minority-White-elected President. With only 39 percent of the White vote, how did Clinton win an election in which the electorate was 87 percent White? He took the White House with an anti-Republican Black vote that accounted for 18 to 50 percent of Clinton's total in strategic electoral states where the White vote was close. That strategic advantage made it possible for Clinton to become President with the majority of votes in only his home state of Arkansas, less than a quarter of those eligible to vote and only 39 percent of the White vote.

I'll explain it another way. One opinion poll prior to the election showed Clinton and Bush in a statistical tie (44 percent to 43 percent respectively) for the White vote. However, Clinton's entire twelve-point nationwide lead was the result of over 90 percent Black support. Had Bush gotten only 20 percent of the Black vote, he would still be running the country.

No matter what some Republicans personally think of Black people, it's just not smart politics to ignore a significant voting bloc that is strategically placed. But a GOP strategy that goes after the "Black vote" without appealing to distinct segments or niches of those voters is a flawed strategy also.

A Black Strategy for Leveraging Clout: The 35—65 Solution

If the GOP targets Black entrepreneurs and Black self-help organizations with assistance and relief from regulation and taxation, it will get the necessary 20 percent threshold share of the Black vote and help itself to the White House and Congress in the process. That would empower the Republicans, but not the Blacks.

Make no mistake, it would not please me to see Blacks exploited by the Republican Party just as they have been by the Democratic Party. It would only help democratize American politics if Blacks manipulated both major parties by bolting from the stranglehold of the Democratic Party. But I don't want to see them snugly in the arms of the Republicans either. Blacks need self-sufficiency as well as government assistance. And both parties can uniquely support this dual agenda.

In this scenario, Blacks can focus on the two major parties: Democrats for social programs and Republicans for economic self-sufficiency and to guard entrepreneurial activities from an intrusive and tax-happy government. But because the Democratic Party gets about 90 percent of the Black vote just for the asking, there is no incentive to advocate self-help and empowerment and there is no way to make the Democratic Party accountable.

Blacks can change that. The strategy for doing this is taken from another history lesson. In the future, the ability of the race-based "Southern strategy" to win White votes is actually in the hands of Black voters, not of White Republicans. Blacks, properly schooled in the history of the past relationship between themselves and both parties, can end the Southern strategy and Democratic negligence at the same time. They can do it with a two-party or even three-party political strategy, while keeping in mind that there is only so much to be gotten from the political system under any arrangement.

It's been done once before by Blacks themselves. I call it the *35-65 Solution*. Blacks can solve this dilemma by returning to the voting split they utilized between 1936 and 1964. In that period, Blacks voted roughly 65 percent Democratic to 35 percent Republican. Both parties had to compete for their votes.[14]

During that period, both parties fell over each other passing bills that benefited Blacks because—and this is the point—Blacks *strategically* used their voting power. That means that if Blacks should split their votes today—35 percent voting for the Republicans and 65 percent for the Democrats—the partisan competition of both parties for the Black vote would break the political logjam now blocking Black economic and social advancement. In this case, the 35-65 Solution will end the possibility of domination and exploitation of the Black community by either party.

Most significant, the political demands of Blacks would be shaped by this strategy. Instead of the current 100 percent noneconomic socialist agendas, Blacks would insist on wealth creation and self-empowerment. Black Republicanism would bridge the chasm between Black deprivation and Black success. Instead of the BUM, Blacks would have an accountable leadership—Democratic and Republican.

With a 35-65 Solution, Blacks would also be represented by a Republican Congressional Black Caucus that would emphasize economic self-sufficiency and training and education to prepare all Blacks for a twenty-first-century high-tech workforce. The goal: Make the middle class as large as possi-

ble and the underclass as small as possible, and let the rich
Black people pay their own bills and compete with the rest
of the American population without hogging the resources
needed by the Black young, the working poor, and the un-
derclass. Until such a body is elected, the current or future
Black Republicans in Congress should organize one con-
sisting of an advisory body, if necessary. Interestingly, a 1992
HBO/Joint Center Survey fund that 33 percent of voting-age
Blacks self-identify ideologically as conservative, 30 percent
claim to be moderate, and 29 percent see themselves as lib-
eral. Despite the fact that the survey's author, criticized by
"my liberal friends" for asking the question and fearful that
conservatives would use his findings as propaganda fodder,
interprets the responses as having little political significance,
they reinforce my belief that the philosophical ideals (not
political objectives) of most Blacks and the Republican Party
are identical.[15] However, Blacks have been turned off by the
recent racial conduct of Republicans, and, conversely, Re-
publicans have been turned off by an endemic socialism
among Blacks. I share both concerns. Black people who are
interested in economic growth need more legislative relief
and less grandstanding.

Three U.S. Supreme Court decisions in June 1995 have
imposed strict new limits on three federal race-based prefer-
ence programs—affirmative action, busing for school assimi-
lation, and race-based set-asides for Congress—to combat the
effects of discrimination by Whites. Two clear-cut messages
became obvious once again: (1) These preferential tokens
are inadequate and failed attempts to attain parity for Blacks
and Whites; and (2) the White majority, speaking through
the highest court in the land, has said once more that it has
almost given up on Black America.

Jesse Jackson's predictable response to this triple hit was

to attack his political enemy, the only Black on the Supreme Court. He suggested that Clarence Thomas is a traitor who wants his own people to be returned to the status of chattel "slaves." This typical Black leadership reaction, aside from its unconscionable absurdity and misreading of history, is inept and an obvious grandstanding gesture. The truth is that professional Black leaders have no plan to plan an alternative to dependence on White paternalism that will achieve political, educational, and economic parity for Blacks with other Americans, discrimination notwithstanding.

If, after almost four hundred uninterrupted years as chattel slaves and marginalized victims of White racism, Jim Crow, state-sponsored segregation, prejudice, and not-too-subtle discrimination, Blacks don't recognize the necessity of self-organization, they are indeed inferior—at least in judgment. My judgment tells me that while the courts and the White majority are debating and constantly reversing their own decisions on the question of fundemental rights of the rapidly deteriorating Black minority, Blacks should not depend on most White people, the BUM, or the government.

"It never ceases to amaze me that the courts are so willing to assume that anything that is predominantly black must be inferior," Justice Thomas wrote in defending the inherent equality of Black institutions and Black people—even when separated from Whites. This scathing upbraiding of liberal racism followed the high court's 1995 ruling against a plan of one-way integrationism that spent $15 billion—not to educate Black students—to get suburban Whites to sit next to them.[16] The theory that Blacks cannot learn unless accompanied by Whites is "a jurisprudence based upon a theory of black inferiority,"[17] Justice Thomas added.

After fifty years of *Brown*, it should be obvious that a Black student needs an education more than a White friend. This is the knock-down punch that makes Jackson's liberal-racist policies—not Thomas's advocacy of racial equality—stand out as pro-slavery, *i.e.*, oppression by a grinding poverty. Since 1954 and *Brown*, Black and White liberals have mounted a movement to validate Blacks as a biologically inferior group in order to achieve parity with Whites. In utilizing that strategy to achieve that goal in a society of endless discrimination against them, Blacks must eventually certify themselves as inherently inferior and therefore eligible for lifelong preferences.

The only alternative is for Blacks to become self-sufficient by competing politically, educationally, and economically with Whites—while living with racism and discrimination. The first step is self-realization—to psychologically let White people go; the next step for Blacks in achieving parity is to recognize complete reliance on a mercurial government or White paternalism as counterproductive; and, finally, to start fresh with a new paradigm of self-empowerment.

CHAPTER 11

A PLAN TO MAKE
BLACK AMERICA
WORK

Up! Up! You mighty race. You can accomplish what you will.
—MARCUS GARVEY

It was a day of celebration when Rick Singletary opened the largest Black-owned supermarket in the country in Columbus, Ohio—a spectacular $4.4 million operation. He had worked for a major grocery chain for fourteen years and started his own store with his life savings, those of his mother, and a government-insured loan from the Reagan administration. He located Singletary Plaza Mart in the Black community because he knew there was a need for a grocery store there, and because he wanted to create jobs for Blacks.

The entrepreneur needed only a $200,000-a-week volume to keep 130 Black people working. And yet, in a tragedy that exemplifies the real reason why Black America has never been able to compete with White America, Singletary's store failed. Although his research had shown that Blacks in Columbus spent $2.5 million per week on groceries, he could not get them to spend even $200,000 of it in the store he had built for them in their own neighborhood.

I am familiar with the details because I tried to help Singletary, and I tried to help the Blacks in his community realize what was happening. For three days, I joined others in the Buy Freedom campaign of Black economic empowerment in Columbus. But, sadly, we failed to save his store.

This is not simply a neighborhood issue, it is a national disgrace. Rick Singletary, a good man who banked on his community, went bankrupt. He lost his life savings and his mother's savings, and 130 Black people lost their jobs. *This story is repeated somewhere in the Black community every day.* This gives credence to my theory that the most successful economic boycott ever conducted in America is the boycott by Blacks of their own businesses.

And yet, I have been attacked in a Black New York weekly, the *Amsterdam News,* for my work in promoting self-help among Blacks. It is obvious from the social and economic crisis in the Black community that most Blacks— certainly not the leaders—see no need for self-reliance—even in the face of a 1995 Supreme Court decision that severely cripples affirmative action programs. So in this chapter I will make my case for self-determination and freedom from liberal White racist hegemony.

The reparations package advanced by Black leaders includes various race-based entitlement programs called affirmative action. The purpose is to establish parity in business by "setting aside" lucrative contracts; reducing the academic standards for entrance to college for all Blacks (as is now the practice for some Whites); and recruiting and promoting Blacks in the corporate sector as a gesture of goodwill. Grudgingly over the past twenty-five years, White America has paid billions of dollars to atone for past sins and to buy back its soul from Thomas Jefferson's "Avenging Deity"

Instead, after a quarter century of ignoring disadvantaged Blacks and redirecting an estimated 80 percent of affirmative action preferences to middle-class White women as well as middle-class Blacks and Hispanics, this laudable albeit clumsy attempt to right past wrongs has been undercut by an ignorance of how to help. Billions of dollars later, we ended up stabilizing poverty, illiteracy, and the "disadvantages" the policy intended to obliterate. The unemployment rate for out-of-work Blacks is still twice that of Whites, and the proportion of Blacks in the private sector and in higher education has not equaled the percentage of Blacks in the population, 12 percent. We are losing in the streets, and in the classroom also. In 1977, Asians on temporary visas in this country earned approximately the same number of Ph.D.s as did Black Americans. Fifteen years later, Ph.D.s awarded to Asians on temporary visas has soared to 6,464— nearly seven times more. In that same year, only 951 Ph.D.s were earned by Black Americans.

Although the Talented Tenth has ripped off most of the entitlements that were designed to aid the socially and economically disadvantaged in their community, the misappropriation of the Small Business Administration's 8(a) set-aside program is particularly egregious. Of the $4.4 billion dispersed in contracts, 25 percent went to a tony 1 percent of the businesses. That sets a new BUM standard for exploiting the Black poor. But it is by no means a new practice.

The Tulsa Riot: History Repeats Itself

Often, Blacks have it within their power to determine their own destiny and to become economically empowered, but they often walk away from the opportunity and fall victim to the entitlement mentality in which they seek acceptance from Whites rather than equality with them.

What happened in the Greenwood neighborhood of Tulsa, Oklahoma, offers a history lesson that Blacks need to learn in order to take advantage of opportunities. This is an example of how Blacks can succeed when they are self-reliant, and also of how they fail when they become self-deluded.

In the early 1900s, the Black neighborhood in Tulsa had produced some of the richest Black families in America through economic development and entrepreneurship. In Greenwood, Black Tulsans created what became known as "the Negro Wall Street," a financial district so prosperous that it could serve as a model today. Blacks owned their own property and operated businesses, hospitals, banks, theaters, hotels, and schools.

Then, on the last day of May in 1921, there was a race riot in Tulsa that destroyed this Black economic center. What really happened that day is difficult to determine because of cover-ups at the time and since. Some Whites claimed that the riot began when a White woman was allegedly attacked in a downtown elevator by a nineteen-year-old shoeshine boy named Dick Roland. He was never officially charged, but a lynching party was formed. In fact, a headline in the *Tulsa Tribune* allegedly said TO LYNCH NEGRO TONIGHT. I say "allegedly" because copies of the article and its headline have been torn out of the archive editions of that newspaper.

269

Many believe that what happened in Tulsa actually had little to do with the charge of rape. It is widely suspected that many Whites had been waiting for an excuse to destroy the Black economic base in Tulsa, and that is what they did that night. A White mob paraded through the streets with high-powered weapons, hunting down Blacks. They crossed the tracks into the Black business district and began to loot and burn.

By the next morning, Tulsans awakened to the sight of nearly the entire thirty-six-block Greenwood neighborhood ablaze. In one reported incident, five Black men were trapped in a burning house, and when one tried to escape, he was shot by a White man and his body was then thrown back into the flames.

While it was never officially confirmed or acknowledged, many Blacks reported seeing an airplane drop what appeared to be fire bombs on their buildings. Others recalled being shot at from the air. If this is true, it puts Tulsa in the history books as the first American city to be intentionally bombed from the air. The official death count was thirty-six, but some reports put the number at ten times that. This act of cowardice by hate-mongers precursed the 1995 bombing of the federal building in Oklahoma.

In the months and years that followed, Black businessmen did try to rebuild their economic base in Greenwood, but they ran up against the White power brokers at every turn. Attempts to finance new buildings through loans or donations were blocked. An ordinance was passed to stop Blacks from rebuilding their homes. During that winter, more than a thousand Blacks were forced to live in tents. Later, Walter White of the NAACP believed that the ordinance was the work of White businessmen who had been trying to buy the Greenwood land for years in order to turn it into an industrial complex.

But the Blacks of Tulsa persevered, and, to their credit, they rebuilt Greenwood into an even more prosperous community. Unfortunately the community that survived a racially motivated attack by Whites did not survive another, far more insidious assault on Black self-sufficiency—integration. When Whites relaxed the Jim Crow laws, Blacks voluntarily walked away from businesses owned by their own people to spend their money in White business that previously had not let them in the door. In turning away from Black businesses, they accomplished what racism and the riot had failed to do—they destroyed their own community.

Making Blacks Competitive

The key to making Black America competitive with White America is really quite simple. Black Americans now earn nearly $500 billion annually, according to economist Andrew F. Brommer.[1] This is roughly equivalent to the gross domestic product of Canada or Australia. And yet Blacks spend only 3 percent of their income with a Black business or Black professional.[2] By spending 97 percent of their money outside of their racial community, they exacerbate their own social and economic problems.

This is the reason that Blacks do not keep pace economically or socially with the rest of the country. Since 80 percent of Americans are employed in small businesses, it is common sense that if businesses in the Black neighborhoods do not flourish, job opportunities will be greatly reduced.

To succeed as a people, Blacks have to invest in and

build their community. Other ethnic groups turn their money over multiple times within their communities. If money turns over ten times, it means that for every $100 spent by an individual, nine other individuals or businesses will have access to that same $100. This investment increases the community's economic strength by $1,000 instead of just $100.

It works this way. You earn $100 a week and I earn $100 a week. You give me ninety-seven of your dollars. I'm living on $197 and you're living on $3. How can your house be as big as mine? How can your car be as new as mine? How, even, can your IQ be as high as mine? Income affects nearly all aspects of life. A higher paycheck means you can afford to live in a better neighborhood with better schools and more opportunities for intellectual development. Studies have found that the group in America with the highest income is the group with the highest IQ. The group with the second-highest income is the group with the second-highest IQ. The overall IQ of Blacks is low in part because the income retained by Blacks is at the bottom.

Take Back Your Mind

Rick Singletary knows this all too well. The problem is not that Blacks don't have money. The problem is what we do with it, or don't do with it. Just as we waste our votes by not demanding anything in return, we don't spend our money where it pays off.

Over the last twenty-five years, the Black community has had a major thrust in politics and civil rights. We have staged Freedom Marches, but we have never stopped to think about what really buys freedom. It isn't worn-out shoes, and it isn't even civil rights legislation. True freedom springs from economic parity with other Americans.

Money is not everything, but I rate it right up there with oxygen. After almost one hundred years of social engineering, Blacks can sit next to White people in classrooms and restaurants and on airplanes, but can they afford it? *The bottom line is that the only color of freedom is green.* Pride, education, and economic self-sufficiency were the message of Marcus Garvey and Booker T. Washington. But those two great Black men were vilified by the self-serving, self-hating elitists among their own people, and their vital message of self-reliance was blocked. Instead Blacks have spent decades with their arms extended and their hands out, doing the economic death dance to the tune of integration.

To succeed in attaining parity with Whites, Blacks must be psychologically capable of competing with Whites and other Americans. After centuries of exposure to White supremacy, scientific racism, and government policies that inspire racial self-hatred, Black self-doubt is still a growth industry—one that is perhaps second to none, including Black imprisonment.

This self-doubt has been further reinforced by "race norming," the practice of holding Blacks to a *colored standard* of performance while Whites are held to the *highest standard* of performance to earn the same benefit. The insidious psychological climate that White liberal racism has created for young Blacks—a milieu in which self-doubt is reinforced by White hostility and resentment over "preference received"—produces unimaginable stress and subsequent failure. During legal segregation, Blacks were not so intent on collecting righteous rewards. Instead, they concentrated on excellence and successful performance, which ultimately boosted self-esteem and led to professional achievement. Blacks performed as well as Whites when given an equal background and opportunity. This differs markedly from the contemporary scene in which Blacks with comparable socioeconomic status perform academically below their White peers.

Racism is not the primary reason Blacks are at the bottom of the societal ladder. Blacks are subordinated primarily because of their own lack of economic organization. This chapter explains how it happened and offers an explanation for why it happened.

Reluctant Entrepreneurs

The Black community's socioeconomic failure is not an accident. The Black community was never organized to become economically self-sufficient because of a cruel plan of non-economic socialism that was imposed on the Black community by Black and White liberals, many of whom were socialists. This plan was carried out through the original sin of integration (assimilation) that guaranteed and institutionalized Black failure in the United States.

Blacks have been reenslaved as "the worst victims of the economics of scarcity,"[3] by the liberal White racists and by a Black Unaccountable Machine that has historically had its hand out for entitlements rather than having its hands on the wheel of self-reliance.

Other ethnic and racial groups continually overcome America's racist predisposition with free-market tactics—racism, anti-Semitism, and religious bias notwithstanding.

Author Joel Kotkin was being kind when he referred to native-born Black Americans as "reluctant entrepreneurs." When Kotkin compared the median income of families of Caribbean ancestry in the United States, he found that it exceeded the level of White Americans: "The West Indian experience is the best indication that it is historic and cultural factors, not skin color, that best explain the disinclination toward business among so much of Black America—a disinclination that borders on conviction."[4]

This "disinclination" arises from the fact that socialism is embedded deeply in the psyche of our American Black leadership, which has perpetuated a slave heritage instead of calling upon our proud African background as a basis of productivity. Ironically, while Black Americans ask "Where's mine?" Blacks of Caribbean descent have gotten theirs by embracing cooperative traditions, just as Chinese, Japanese, West African, and Korean immigrants have done.

Members of these racial groups have contradicted the two-hundred-year-old explanation that native-born Black Americans don't own businesses because racially prejudiced lenders won't provide them with financing. This explanation of Black economic failure is so ingrained among Blacks and so apparently rational given America's racist history that it is rarely, if ever challenged.

Several participants blamed White racism for impeding Black economic development at the Dow Jones–*Wall Street Journal* conference in 1993. One rousing speech on the need for more liberal lending by banks was given a standing ovation.[5]

The next year, I delivered the keynote address at the same conference's meeting in Baltimore. Tired of the lies and misperceptions on both sides of the issue, Black and

White, I was intent on dispensing some truth. But I was also aware that I was about to dispute theories of blame that had been embraced for nearly two hundred years. I knew my words might not be well received, but I had the truth on my side.

In *Ethnic Enterprise in America*, author Ivan Light asserts that the classic discrimination-in-lending theory as the preeminent explanation for Black business retardation has lost its force of logic. He notes that studies of small businessmen have shown that, contrary to expectation, institutional loans have been relatively insignificant as a financial resource tapped by entrepreneurs. Only a small percentage of business people have reported seeking or obtaining bank loans in order to open a small business, Light found. A far greater percentage rely entirely on their personal resources, especially their own savings, and loans from family and friends. "Since bank credit has been so insignificant a resource for new proprietors in general, even complete denial of bank credit could hardly account for the Negro's singular difficulties in small business,"[6] Light noted.

The Black Economic Tradition

Backed by truth and Light, I suggested to the Dow Jones–*Wall Street Journal* conference that Blacks take control of their own economic development by pooling their resources and leveraging financial clout.

In my address at the conference, I noted also that Black entrepreneurs could finance their ventures by saving and borrowing against their own equity. I implored Blacks to call on their own culture for the answer. I pointed to the tradition of the Yoruba *esusu*, the African informal lending associations in which individual income is pooled in order to leverage capital more efficiently. The Japanese, Koreans, and other ethnic and racial groups capitalize small businesses in this manner.

It has become painfully obvious that Blacks are unaware of their own tradition of economic self-help. African slaves in the United States were abducted from West Africa, where tribal credit associations were widespread. Today Black Americans—the descendants of West Africans—are just about the only people in the world who no longer use this informal financial institution from their own culture. Scholars say the *esusu* custom is indigenous to Africa and developed in southeastern Nigeria among the Yoruba people. Ironically, it was the slaves from the West Indies who embraced the custom when they learned of it from their African brothers and sisters, and Blacks from the Caribbean are credited with keeping the tradition alive.

When West Indians began migrating to the United States around 1900, they brought the *esusu* tradition with them. Some native-born Black Americans called them "stingy" and resented them then and even today because of their thrift and competitiveness and, perhaps, a whiff of arrogance. Whatever the source of conflict, a difference in attitude and the ability to start and manage small businesses was apparent. Not surprisingly, one of Black America's leading proponents of entrepreneurial and economic development, *Black Enterprise* magazine, is the creation of Earl Graves, whose parents are from Barbados in the Caribbean.

West Indians filled the ranks of Marcus Garvey's economic self-help mass movement, the Universal Negro Improvement Association, the largest among Blacks in history.

They were aggressive in starting retail businesses in direct competition with Whites—grocery stores, tailor shops, jewelry stores, fruit vending, real estate, and so on—while native-born Blacks resented them and followed a noneconomic policy of seeking government entitlements rather than trying to build their own economic-support networks.[7]

Focusing on the Economic Solutions

Blacks need to get off the racial problem and onto the economic solution. Black leadership has rarely stressed solutions in modern times. The major exception among integrationists was Martin Luther King, Jr., who advocated pluralism and economic self-development. In a theme that has been adopted by Nelson Mandela in South Africa, King told Blacks, "New laws are not enough. The emergency we now face is economic."[8]

Black flight from our cultural foundation resulted in the surrender of our schools, our businesses, our self-employment, the future of our children and families—essentially our destiny. We lost all this in trying to become integrated and assimilated into White centers of power. It never happened. In truth, we were naive about power. Blacks confused being accepted by Whites with being equal to them. Meanwhile, other ethnic groups ignore both assimilation and the melting-pot myth. Their overachievement demonstrates that group unity is simply more powerful than racist opposition.

In a special issue on immigrants, *Time* magazine surveyed the landscape of newcomers and assessed their Americanization. "With a mixture of animosity and admiration, a long-impoverished minority watches as a new wave leaves them behind,"[9] *Time* commented on Blacks.

The magazine noted that since the first slave ships arrived in America, Blacks have watched in anger and humiliation as wave after wave of new immigrants surpassed them. In Harlem, many Blacks call for running Koreans out; in Miami, some blame the Cubans and Haitians for working for minimum wage; and in Houston, where the Indochinese run mom-and-pop stores in Black neighborhoods, Blacks buy in wonderment at the efficiency of these Asians.

Blaming others for exploiting the free marketplace puts no meat or potatoes on your table. The problem is with us, not them. Cultural unity and economic development will free Blacks. Racism, discrimination, and a debilitating welfare system have caused Blacks to lose faith in themselves; Blacks will improve their communities only when they regain that faith. God and economics—or faith and money—are an unbeatable combination.

One race of people cannot expect another to do for it what it refuses to do for itself. Blacks cannot spend 97 percent of their money with non-Black people and blame them for 100 percent of their problems. A San Francisco study showed that money changes hands from five to six times in the Chinese community before it leaves, four to five times among Jews, and three to four times among Japanese. Blacks retain money less than four hours before it leaves the Black community.

If Blacks spent 50 percent of their disposable income with Black businesses and targeted the other 50 percent for only non-Black firms that support the training and education of Blacks, the market economy would produce the human and financial capital that would overcome racism and economic marginality. It would also close the inequality gap in median family income with Whites—something affirmative action can never do.

There are thirty million of us that the Census Bureau can find. And although we are only 12 percent of the United States population, 40 percent of all records sold in America are purchased by Black teens. Fifty percent of all tickets to movie theaters in America are purchased by Blacks between twelve and twenty-four. Fifty-two percent of us own our own homes. We are 12 percent of the population and we drink 20 percent of the Scotch whisky. If you took Blacks out of America, Wall Street would collapse "last week," as I like to say. So much of our potential human capital goes undeveloped because far too many of our young people are not provided with the essential tools or the guiding vision to help them become part of the solution. By waiting for others to save us, we let our children become prey.

As if things weren't bad enough, Blacks are marching into the twenty-first-century competition of engineering, computer technology, science, and mathematics unarmed. In spite of the enormous opportunities that cultural diversity will create in the next century, Blacks could be left out, not by discrimination, but by the lack of preparation. One-third of Blacks have not even finished high school.

Black Brainpower

Racists have always tried to cast Blacks as intellectually inferior in order to justify enslaving or otherwise marginalizing them. Racist philosopher David Hume claimed that Blacks are naturally inferior to whites. One of the most recent written attacks on Black intellect, *The Bell Curve*, claims to offer scientific proof that Blacks are intellectually inferior because of their genetic makeup. It is but another soothing bedtime story for the dedicated racists.

I have some material that might disturb their dreams. There is now empirical evidence that should settle the argument regarding the alleged and often presumed intellectual inferiority of Africans. The 19th Decennial Census of the Population of Great Britain report in 1991 may have put it to bed for good.

Theodore Cross broke a story in the *Journal of Blacks in Higher Education* that went largely unnoticed in the British press and was completely ignored by the American media. It was headlined "Black Africans Now the Most Highly Educated Group in British Society." The story noted that over 26 percent of adult Black Africans, compared to only 13.4 percent of White adults, in the United Kingdom hold academic qualifications higher than "A" or college level. Thus, Blacks in the United Kingdom with African origins are outperforming British Whites by two to one.[10]

Asians who are dominating the American education scene and swamping Blacks in achievement and the Caribbean Blacks who best Whites in the United States take a backseat to the Africans in the United Kingdom, where race conflicts are even more contentious and where the academic and scholarly cultures are "centerpieces of Western Eurocentrism." Cross notes also that African Blacks in Great Britain are "almost twice as likely as British Whites to hold positions such as barristers and doctors, or to be employed in the other professions."[11]

In his article, Cross addresses racists and the White liberal racists who have advocated dependency rather than self-reliance for Black Americans. "The powerful performance of

Africans in Britain puts a solid nail in the coffin of scientific racism, a thesis that holds blacks have intellectual inferiority encoded in their genes," he notes. "Further, the striking success of African blacks in the U.K. should be deeply challenging to the views of American liberals who support racial integration and equal opportunity for blacks but who nevertheless believe that Negroes are a debased and inferior minority."[12]

Cross summarizes, "Clearly, the British experience shows that blacks do not suffer from an inherent disadvantage that will cause them to fail. It is high time to admire blacks for their brains rather than their brawn."[13]

The superior performance of Africans demonstrates just how much White liberal paternalism, integrationism, and assimilationism have restricted Black Americans. On the other hand, it gives hope because it shows the power and the potential of having African blood, kinky hair, and big noses. Most of all, it shows those defeatists among Black Americans, who profess that "education is Whitey's thing," that they are a corruption of a culture. Hip-hopping, jive-assing, excuse-making, Jew-hating, crabs-in-the-barreling, joke-telling, play-acting, rhyme-making, and culture-killing—these are not the true traits of a proud people. Blacks have learned to feel inferior from the low expectations that others have of them.

A Plan to Make Black America Work

I don't know what White people owe us, but I am convinced they are not going to pay us. As Billie Holliday sang, "God bless the child that's got his own."

I propose that rather than blaming others for their failures, Blacks focus on solutions, particularly on education and training and small business development. I suggest setting up a partnership between government and a Black Leadership Network (BLN).

This network would distinguish itself from the BUM in two ways. It would have a strategic twenty-year plan divided into three phases: Planning, Implementation, and Evaluation. This new leadership would become a working group divided into strategic and operational functions supported by Black professionals with real-world expertise.

By pooling its financial resources, the Black Leadership Network would provide sufficient capital to launch this human resources reclamation project. It would be directly responsible for funding all three phases. I predict that if the rest of the country saw the Black community exercise this kind of leadership and initiative, all forms of support would be forthcoming. If so, I would recommend that governments at city, county, state, and federal levels match the BLN's funds on a nonpartisan basis.

If thirty-two million people earning as much money annually as the gross domestic product of Canada or Australia cannot govern themselves and finance their own empowerment, they will never have equality and it cannot be conferred upon them. It is not unreasonable to expect Black organizations to put up a minimum of $5 million to get started. If they do not have it, how is it that the same 350 Black organizations spend $16 billion a year[14] attending annual meetings? How does the Talented Tenth spend $200 million to $500 million every September at the Democratic Congressional Black Caucus's five-day party?[15]

For purposes of this discussion, let us assume that these very well educated and affluent people do not have any money. The answer to that objection is that they can use the American market economy to get it. That is what the Girl Scouts do: They earn millions of dollars a year by selling cookies to fund their programs. If one million Girl Scouts can sell cookies to finance the growth of young women,[16] why can't thirty-two million Black people sell a product or service and buy their freedom? If Blacks do not learn to lead themselves, someone else will always lead them—and outside leadership will always be to their detriment.

Project U.S. is my plan to reinvent the Black community. It is fundamentally an economic solution to an economic crisis. Let me offer you a prototype for taking a close-knit virtual community into cyberspace.

PROJECT U.S.: A Proposal to Make Black America Work

1. A Leadership Plan. Black America could organize its leadership into a nationwide network of groups, organizations, and churches and: (a) agree on a twenty-year plan to remake the Black community; (b) accept specific responsibilities for executing the overall plan; and (c) provide the necessary resources.

Many of the leading-edge occupations of the next century will require computer, mathematical, and scientific training, which the Black masses simply have not developed. To overcome this educational gap, Blacks must become competitive in these areas, especially in information technology.

Each organization in the network should accept a specific task in implementing a plan that will close this gap, such as the identification of candidates for education and training, supporting them with emotional and financial assistance and special tutoring, and job placement upon completion of the program.

2. An Economic Shield. Fifty percent of all Chinese-Americans, who have the highest median family incomes of any cultural group, have worked at one time in their lives in a Chinese-owned restaurant.[17] Asian-Indian-Americans focus on motel ownership. Korean-Americans specialize in the grocery and laundromat businesses, and Jewish economic power grew out of the retail and entertainment arenas. Black Americans need to focus their economic development on a business activity—an economic shield that is compatible with their resources, development, and cultural needs. I recommend as their economic foundation any of these ripe areas: electronic marketing on the Internet, long-distance telecommunications services, home-based marketing, and hotel ownership.

Black organizations spend $16 billion a year attending conventions, and Blacks do not own one major hotel in the United States. As keynote speaker at the NAACP's economic summit in 1990, I urged the Black leaders of 150 groups to cancel their respective conventions for one year and use $3 billion of the $16 billion that they spend annually for meetings to purchase hotels that would allow them to establish an economic infrastructure for the Black community.

I proposed that the Talented Tenth use their respective conventions not to celebrate their successes, but to share them. I suggested they convene to develop and coordinate an economic recovery plan—which includes the purchase of

285

hotels in major cities. For example, a hotel in one of the top twenty markets can be purchased for about $200 million with a down payment of 10 percent—or $20 million.

With $400 million (about the annual interest on $3 billion), a Black business group could own first-rate hotels in the twenty largest urban areas, where over 50 percent of the Black population resides. In turn, these hotels would be the recipients of our annual $16 billion convention market. The twenty hotels could generate $1 billion in income annually. These Black-owned hotels could solicit bids from Black-owned caterers and furniture makers and from Black architects and attorneys, among a variety of other small Black firms. Non-Black firms that conduct business with Blacks and support community-service programs would be urged to bid.

By becoming major players in the mainstream economy through the hotel business, Blacks could build the financial resources to deal with the social problems now discussed at their national meetings. Under this plan, Blacks would be in a position to use their pooled economic resources as a capital base. By creating a modern version of the traditional African Yoruba *esusu*, we could finance our own enterprises and jump-start projects to attack the social problems of poor Blacks.

This $3 billion could be leveraged to $30 billion in financial markets and deposited in Black lending institutions. Some of it could be used for loans to self-starting community people who want to develop their own businesses, some to expand existing businesses and as a source of capital for Black entrepreneurial development. With twenty million members, the eighty thousand Black churches can serve as the capital redistribution base. According to government statistics and *Black Enterprise* magazine, Black churches account for 90 percent of all Black giving—a whopping $2 billion annually.[18]

3. *Entrepreneurial Development*. Blacks should be encouraged to enter the entrepreneuial arena at whatever level they feel comfortable. "Circle of friends" marketing programs are entry-level entrepreneurial endeavors that require small initial investments, often under $100. In these businesses, marketing and business skills often develop quickly, and an extra income of only $200 a month can make a big difference both symbolically and practically.

4. *Technical Training*. "A college degree isn't the only tool for cracking today's job market,"[19] the *Wall Street Journal* said in a front-page story. That is the advice of Booker T. Washington come home to haunt those Black elitists who have refused to acknowledge that there are those whose skills can be developed outside the academic track.

The same story reported that "70 percent of jobs in the next decade will require training other than college." The report noted that "high-school graduates who received training to qualify for jobs and to improve their skills had slightly higher median earnings than college graduates who received no such training."[20]

5. *A Reverse Migration: From Urban Ghettos to Rural Job Markets*. We have a permanent underclass of unemployed Blacks who will never be productive as long as they remain mired in urban ghettos. The great Northern cities simply no longer have the resources, let alone the desire, to address the problems of the Black urban underclass. "The dramatic shifts in the nation's job market are working against urban African Americans and some experts worry the trend will only worsen as new job growth continues to flock to the suburbs,"[21] the *Chicago Sun-Times* reported.

Rather than remaining in the ghettos and thereby remaining part of the predicament, Blacks should move up by moving out. I suggest a massive training and education program aimed at preparing urban Blacks for employment opportunities outside the stagnant urban ghetto and in the American hinterlands, where, if you haven't noticed, the economy is flourishing in many areas.

Many of the major employers in this country and other nations have looked to the hills of Kentucky and Tennessee, the prairies of Illinois and Indiana. Inner-city Blacks should recognize this shift and go after those jobs just as their grandparents abandoned the rural ghettos and moved to the cities to take jobs in the industries that then thrived there.

There are thousands of small towns and villages in the United States which could easily absorb a small number of Blacks.

The major thrust of the emigration would not be to White communities, but to the development of "new towns"—cooperatives of a sort along the lines of a modified kibbutz. Former military installations could also be utilized.

The Twenty-First-Century Gold Rush

At the dawn of the television age, in the 1950s, some people were slow to embrace the wide-ranging potential of this new medium of communication, but those with a vision of what television could be eventually profited enormously. The same holds true today when it comes to the great potential of the information superhighway, known as "the Internet," that takes its computer-savvy travelers into the heretofore unexplored regions of cyberspace.

The Internet accounted for $200 million in consumer sales in 1994, which leaped to over $1 billion in one year. Within five years, 25 percent of consumer sales in the United States will be made electronically.[22] All you need to sell a product or service from your World Wide Web storefront is a "home page." You can make a living at home with your PC. This is what will happen to those among the technologically displaced workers who have foresight.

To be sure, a great many people have already realized the potential of the information superhighway. In some ways, the mad rush into cyberspace is reminiscent of the gold rush of 1848. Hordes of fortune hunters, thrill seekers, and information gatherers are already staking claims on the Internet frontier. Nearly fifty million individuals as well as most major organizations and companies in the world have settled in to mine it for all they can. My goal is to make sure Black Americans are among the first to stake a claim.

The Black economic agenda has been dead in the water in this country. But I am blasting it into cyberspace.

Our future lies in this developing frontier. And it is my plan to pioneer this region as a source of wealth and economic vitality for Black Americans. That is why my New York–based radio talk show on WLIB, the city's most powerful radio voice for Blacks, has become home to a vast audience of Blacks eager either to learn or to share their knowledge of computer technology and the Internet.

Shortly after launching the *Tony Brown* show on WLIB

on April 3, 1995, I invited listeners to talk about the need for improved computer literacy among Blacks. I opened this dialogue because I believe that the Internet holds incredible potential as a marketplace for entrepreneurship by Blacks of limited income. I quickly discovered that I was not alone in this belief. My telephone lines lit up with calls from listeners who also see the potential for economic development in the cyberspace marketplace.

tonybrown.com

In response to this show of great interest, I set up an Internet-accessible computer bulletin board service called Tony Brown Online. On August 6, 1995, my systems operator, Jim Cannady, opened up the phone lines to my computer network, and I invited subscribers in New York City to dial 212-869-5555, or those outside the 212 area code to call for a local access number. Whever you live, Tony Brown Online is always a local call. Our e-mail address is tonybrown.com.

The thousands who have responded to my invitation to join the marketplace of the future have discovered that they can send electronic mail to a global network of thirty million to fifty million people. This worldwide network may grow to one hundred million by 1998 via the Internet. But for Blacks to capitalize on the wealth of opportunities presented by this new form of mass communication, we must promote computer literacy among our people

The ability to navitigate the Internet will be critical in the coming decades. Vast opportunities for wealth creation will open up, particularily in the field of electronic marketing via computer networks. Through electronic marketing, anyone with a personal computer can stay at home and make money, whether on the information superhighway or on a local network bulletin board, known as a BBS.

Our service, Tony Brown Online, is like a cable television system with multiple channels. You can go wherever you want and do whatever you want with services such as full Internet access to an estimated thirty to fifty million people worldwide. The network will also offer interactive services such as electronic mail to people all around the world, local and worldwide matchmaking, shopping malls, databases, health pruducts, discussion groups for people who share similar interests, business and employment opportunities, computer training and professional placement culture, travel, sports, books, music, celebrities, a message board, an on-line fax service, private forums for personal and business teleconferencing, plus opinion polls/surveys, file libraries, entertainment (meet a friend at Tony's After Hours), astronumerology predictions, youth activities (The Cyberspace Cadets), and information on where to find money for college. We also developed the Tony Brown Online Computer Training and Placement Service to train and develop professionals separately as an off-line service.

As a communications system, the duo of *Tony Brown* and Tony Brown Online comprise a virtual community:

1. While listening to the discussion and/or waiting to get through, listeners who are on-line can post electronic questions or messages to me when I am on the air with a guest or caller. I am able immediately to integrate this on-line computer input into the on-air radio discussion.

2. Listeners can communicate on-line via computers with other listeners and callers, as well as with me, during the on-the-air session and can continue the dialogue after-

ward in a computer news forum with or without me or in a private on-line teleconferencing chat with a select few on Tony Brown Online. At some level, computer education and economic networking at the Buy Freedom Shopping Mall continue twenty-four hours a day, seven days a week.

3. This on-line activity encourages other computer users to participate in the on-air radio dialogue, in addition to on-line chat. But more important, members of the radio audience who are also on-line via their computers have become explorers of cyberspace.

We have created a new experience, and the opportunity to become computer literate, for a community that must recognize that it is in danger of being left behind in a rapidly changing world where access to information is the most valuable commodity we have. Blacks must build their own virtual community, and if the reaction to Tony Brown Online is any indication, I believe a large segment of our community is not only up to the challenge, but well on its way to embracing it. I think I've caught a wave into cyberspace for Black America.

The Cyberspace Club

My initial one-hour WLIB radio forum on cyberspace grew quickly into what I call The Cyberspace Club. To my knowledge, this is the first attempt to take a close-knit virtual community and empower it with an on-air radio program employing computer education, cyberspace technology, and the interfacing of a multiuser computer system, or bulletin board network.

Cyberspace doesn't physically exist, of course. It's an

intangible place, formed out of the collective consciousness of the people involved. It is a place where we can find our hopes, wishes, and aspirations—and, perhaps, confront our fears. I saw a way to use computer technology as a tool to alleviate inner-city deterioration, and as a method for providing economic opportunities for our community of listeners.

That's why members of The Cyberspace Club, or those with the "broadest bandwidth," so to speak, are so essential to this effort. They come from among the 25 percent of Blacks who use computers regularly at home and at work.[23] On average, people who own computers earn 15 percent more than those who do not. A computer magazine's advertised reader profile reflects an average household income of $87,500. A caller to WLIB said she doubled her teacher's salary after a computer course. Computer-savvy cyberspace pioneers will inevitably direct this country and the world in the quest to explore the future and, if we are lucky, will help us solve our social and economic problems with new technology.

Therefore, identifying the members of this knowledge class (cyberleaders) on WLIB was step number one in this plan of community education and self-reliance. A new audience of computer illiterate but proactive callers had to be attracted also. Even those aware of the importance of computer literacy have been turned off by typical talk show themes. When I began my radio show, my initial objectives were to spread the message of self-reliance, raise the level of debate above demagoguery and meanness, and empower listeners with information. In this new environment, the techie cyberleaders were eager to interface with a less knowledgeable but computer-eager audience seeking computer literacy.

My sense of urgency springs from by belief that within two years the government will close the window of opportunity to budding entrepreneurs attempting to make money on the Internet. Until then, the potential for electronic marketing on the Internet, if it conforms to "netiquette," is unlimited because it is customer-to-customer driven.

Money and Love on the Internet

For most subscribers to Tony Brown Online, the service represents their first contact with cybermarketing techniques and information technology. It is also their first connection with the vast market offered by this future byway of commerce. While we talk on the radio in New York, Cincinnati, or Chicago, listeners can go on-line with their computers and shop in Paris or Harlem. They can use their web browser to find the hyperlinks in texts on the World Wide Web, which enables them to jump all over the world gathering information while seeing how people make money on the free-wheeling Internet.

Those who want to be more proactive can operate their own independent storefront on the section of the Internet called the World Wide Web, where various "sites" or "home pages" advertise products and services. To help people understand the subtleties of the information superhighway, we conduct seminars that provide the details of this electronic marketing and how to start an on-line business.

In a way, Tony Brown Online is the electronic continuation of the Buy Freedom campaign. This effort to inspire Blacks to take responsibility for their own economic development failed when we tried to save Rick Singletary's store in 1985 (see the beginning of this chapter for details). It also fell short in its effort to convince Blacks to spend more money within their own racial community. We didn't have cyberspace technology then. We do now.

The Buy Freedom campaign now has Tony Brown On-line linked to *Tony Brown* syndicated on radio stations in the urban markets that contain the majority of the nation's thirty-two-million, $500-billion Black consumer market. We are also able to offer several home-based business opportunities to sell direct, retail, or distribute vitamins, health personal care products, and home security products. Another item we can offer is telephone debit cards ("telecards"), which, along with credit cards and checks, can be used as a form of payment for charges. They offer discounts of up to 50 percent on long-distance calls.

On Tony Brown Online, individuals can open a store in our Buy Freedom Shopping Mall to service those who are looking for Black- and Hispanic-owned firms. These entrepreneurs can also operate within the more diverse general business community on The Shopping Mall. Or if their interest is more personal, they may want to use an Internet calling card to attract a soul mate or a potential employer—or both—by using a "site" or "home page."

http://www.tonybrown.com

Cyber–pen pals and marriage proposals as well as job offers are likely to develop on this network. This mixture of text, graphics, film, and video is really a glorified personal ad—a personalized billboard—complete with on-line photographs and extensive background information about the participants. If you are on-line and you have a web browser to the World Wide Web, go to the Tony Brown Home Page at http://www.tonybrown.com. You'll see what I mean.

If you are having difficulty understanding this section of the book, you are on the verge of being left behind. While cyberspace has leveled the playing field for knowledge and education, the anti-technology culture of Blacks is a socio-economic disaster waiting to explode amid the widespread computer literacy of a predominantly White knowledge sector. Even Black college students are largely computer illiterate.

Conspicuous among the computer illiterate and absent from full participation in this electronic era are the Hispanic and Black middle and upper classes that can afford personal computers.[24] It has been suggested by others that the Hispanic disparity with Whites can be explained by the language barrier. I would blame the disparity among affluent Blacks on the guiding tenet among most of the Talented Tenth. This holds that civil rights and affirmative action—rather than economic self-reliance through the exploration of new technology—will create an egalitarian society. The longer they procrastinate and remain AWOL from cyberspace, the more difficult it will be to catch up. While the White middle class has blasted off into cyberspace, middle-class Blacks are Earth-bound discussing affirmative action. If you spend a month learning about cyberspace, you're already a month behind.

Cybertalk: Why We Won't Get Left Out

Support and enthusiasm expressed by my radio audience suggest that we have tapped into a new wave of broadcast communication, and perhaps we have also found the solution to the Black predicament. Our system of communication via the information superhighway and the WLIB radio band can provide the guidance needed for thousands of economically disadvantaged and poorly educated people. It can help them to become self-sufficient. It can encourage and assist those in the Talented Tenth who are anxious about being left out of the cyber–gold rush but are afraid of dipping a toe in the frigid Internet waters.

This phenomenon has a third element you must experience personally to truly comprehend. It is a form of energy— an energy of hope born of finding a way to finally join the economic mainstream. The future is being constructed today, and our radio audience and members of our computer network are working as a team. Against considerable odds, the WLIB family in New York and our extended family across the country are not being left out of the electronic revolution or cyberspace.

"Up! Up! You mighty race. You can accomplish what you will," roared Marcus Garvey.

CHAPTER 12

IF I WERE PRESIDENT, HOW I WOULD MAKE AMERICA WORK FOR U.S.

One man with courage makes a majority.
—ANDREW JACKSON

With the national debt projected to grow by more than $1.3 trillion by 1997, we are no longer living in the United States of America. It's more like the United States of Denial. Having plunged in status from the world's greatest lender to the world's greatest debtor, this country needs a leader for the twenty-first-century who can bring the nation and its people back to reality. No Black lies. No White lies. Only the truth. Now wouldn't that make a good campaign slogan?

I'd like to make a case for a new kind of President of the United States, someone who is not a professional politician, someone devoted to this country, not his or her own egocentric goals. We need a leader who will confront racism, deal decisively with the national debt and the technological displacement of workers, and inspire the rebirth of a failing national characteristic—civic virtue.

I'd like to nominate myself for that job. Why not? For the purpose of offering my solutions to our national problems, I ask the reader to play along with me and assume that I am interested in becoming the President of the United States. Who knows? We might both decide it's not such a bad idea.

The theme for my presidential campaign and my administration would be Team America. Shared sacrifice and change are the keys to a renewal of the country. Whether we realize it or not, the job description for the President in this country has shifted. The Presidency no longer requires someone who can play Cold War games with the Communists. That game is over. Our national leader now faces the very serious business of revitalizing a nation that has slipped badly from global leader to an also-ran, from proud lender to debtor, and from decisive champion of the oppressed to hand-wringing fumbler.

Clinton's Failed Presidency

President Bill Clinton's performance has set new standards for ineptness, and it has resulted in devastation for his Democratic Party as well as stagnation for the country. One study found that after he was inaugurated in January 1993, an optimistic public felt two to one that the country was going in the right direction.[1] Halfway through Clinton's term, the public discovered that the man from Hope, Arkansas, was at the wheel without a map or a compass. One year later, by the same two-to-one margin, Americans felt that the country was going in the wrong direction.

With the Republican sweep of November 8, 1994, elections, I believe the public delivered its judgment. By throwing out a forty-year-old Democratic majority in Congress, the voters called for leadership responsive to an overburdened electorate and a government that will be responsible for the nation's economic growth.

Bankruptcy Future?

Author and former economics editor of the *Wall Street Journal* Alfred L. Malabre, Jr., writes of the "inevitable predicament" that will bring "no easy or painless solution" to our nation's economic plight.[2] His choice of scenarios is default. Malabre envisions the United States government effectively erasing the deficit by forcing creditors—holders of government bonds, bills or notes—to roll over the debt or take a percentage of the face value. Interest payments would also be capped in this "repudiation of a large part of the federal debt." In effect, the government will become an authoritarian body in a totalitarian state, according to his theory. "Our main matter of concern, then, isn't hyperinflation or deflation but the prospect of greatly increased governmental control over all aspects of economic activity," he writes. "This is what the future seems to hold."[3]

Our country is in decline not simply because of the potential of authoritarion rule. It lacks a moral base and a political focus. During the 1992 campaign, Bush was not paying attention, so he lost to Clinton, who apparently can't decide where to focus his attention.

An astute Black Republican President wi
America spirit like myself would not have eith
deficit. Of course, rumor and historian J. A. Rogers have
that we have already had five Black Presidents. So I wouldn't
necessarily be the first. But I'd be a lot better administrator
than "brother" Warren G. Harding and hopefully as great as
Abraham Lincoln.

The Moral Imperative

Another unique asset I would bring to this office is a knowl-
edge of how to "play the game." In *The Game of Life and
How to Play It,* Florence Scovel Schinn explains the rules
with great clarity: "Most people consider life a battle, but it
is not a battle, it is a game" of giving and receiving and the
rules are found in the Old and the New Testaments.[4] The
"game" is simple. What our nation gives, it will receive.

In the opening chapter, I noted the influence that my
caring family and community had on me and how this "no-
blesse oblige" spirit stuck with me from my childhood train-
ing. In this concluding chapter, I would like to focus on the
inner meaning of that concept found in the Bible's Parable
of the Talents. I believe it has great relevance to our na-
tional renewal.

You will remember that I said noblesse oblige means
the moral obligation that we all have to assist anyone in
need. This idea is not a religious concept because many athe-
ists and agnostics practice the same moral virtues of good
character, while many religious adherents do not.

Although my example of spiritual instruction is taken
from the Holy Bible of the Christians, Jews and Muslims

share these teachings also. Many Jewish institutions, for example, have an outstanding legacy of philanthropy, especially toward Blacks. Christianity, Judaism, and Islam all teach their believers to make use of their talents and to improve the general welfare by multiplying the talents of others with charity. America is certainly a nation that is predisposed to operate morally, notwithstanding our constitutional obligation for the separation of church and state. In short, we need to help one another. The United States of Camelot.

"Talents"

In addition to teaching the lesson of the Prodigal Son, Jesus also taught his disciples the Parable of the Talents. Earlier, I used the Parable of the Prodigal Son to teach the meaning of affirmative action; now I will explain how the Parable of the Talents informs us of our responsibilities to every other human being—an essential element in national renewal.

Jesus taught the lesson of three servants and their "talents." In His time that word described the monetary currency; today, of course, it refers to our natural, inherited, God-given abilities. In the parable, one servant (of God) received five talents and was told to "multiply." He was entrepreneurial and used the free-market economy by selling camels to multiply the five into ten and realize a profit of 100 percent. A second servant received two talents. This man lacked the mental agility of the first, but he substituted the "fidelity of toil," hard work, to expand his talents to four. The third servant received one talent and out of fear buried it in a hole in the ground.

The third servant did not multiply. He did not use what

God gave him to help others. He did not understand that his gift (his talents) from God was for "them," not "him." In retaliation, God not only took back that one talent, but He took from the servant everything that he possessed. God gave the third servant's one talent to "him which hath ten talents"—someone who will do God's work of helping others.

The judgment that the one-talent servant received is the same that I feel awaits the BUM—especially its Marxist wing—as well as demagogic politicians. God blessed them as leaders with more talents than the masses because He expected more from them. Instead they have misled His flock, whom God loves dearly. The BUM has buried the hopes and aspirations of the socially and economically disadvantaged. In effect, by their failure to lead, the leaders have become the one-talent servant.

In His displeasure, God takes back the authority to lead and gives it to more virtuous leaders who share His vision of helping others—especially the poor and disadvantaged. God's retribution is swift: "And cast ye the unprofitable servant into outer darkness: there shall be weeping and gnashing of teeth."[5]

Leadership is a service to God. If it fails to enhance the people's well-being, it is corrupt and it will fail in the eyes of the world because it has no other justification. Moreover, God will not only smite a leader who "buries" his talents, He will smite a nation that turns its back on the needy.

The fact that so many Americans do not act or view the world in moral terms is the greatest problem facing America. The American people are losing their ability to distinguish between right and wrong. It becomes more and more apparent with each new generation. That's why a spiritual renewal of the American character is the key to the transformation of

American society. For example, if we know how to tell the difference between right and wrong, we will better understand how to reform welfare, balance the budget, educate our youth, create economic growth, and pass on the moral virtues of character to the next generation.

Like the third servant, the Black community's socialistic oligarchy and the liberal White racists who manipulate it do not believe in the "talents" (or potential) of the poor Black masses. History attests to that fact.

Even in the face of the biblical encouragement to multiply and make a profit, the traditional socialist leanings of the Black leadership oppose capitalism in principle. The American electorate opposes socialism in principle, but is in a full gallop in pursuit of socialist entitlements. As economist Ludwig von Mises mused, there is a lot of support for socialism among its opponents. A free lunch is hard to turn down even when you know in your heart that it will make you sick later on.

A Symbol of What Is Wrong

The infamous Davis-Bacon Act is an unconstitutional throwback to Jim Crow. It was passed by Congress in 1931 and signed into law by a Republican President, Herbert Hoover, in direct violation of the equal protection provision of the Fifth Amendment, with the racist intent of favoring White workers in unions over Black workers on federal construction projects.

The Davis-Bacon Act is anti-taxpayer, anti-Black, anti-capitalist, and a guarantee that the most inefficient worker will be paid the highest wage, which, in turn, guarantees the lowest profit. Just like legal slavery or segregation laws, it will not allow employers to choose their own workers.

In effect, this law creates "set aside" jobs for Whites. In truth, it was created to keep Black unemployed and to tax them for their unemployment. It has turned out to be a quintessential example of entitlement socialism. And it has backfired in the pocketbooks of White taxpayers. The Davis-Bacon Act costs American taxpayers $2 billion in inflated labor costs and more than $100 million in administrative costs annually to small contractors.[6]

It was designed to protect White union workers who could not compete with cheaper, more skilled Black labor, by imposing "prevailing wage" requirements. It was passed explicitly to prevent efficient "cheap, colored labor" from competing with less skilled, more costly White workers. This is government-sanctioned, institutional racism—and what the Davis-Bacon critics call "classic protectionist legislation" for building and construction trade unions. It has contributed to Black and Hispanic marginalization and the perpetuation of material poverty, has increased unemployment, and has fostered an anticapitalist and explicit racism.

Blunting Black Self-reliance

We owe the competitive advantage that Blacks held to the work experience they gained during slavery, and in no small part to the efforts of two ex-slaves after Emancipation. William Hooper Councill organized the Colored Normal School (now Alabama A&M University) in Huntsville, Alabama, in 1875 to teach skilled trades, and Booker T. Washington founded his self-help program at Tuskegee Institute in

Tuskegee, Alabama, in 1881, which had the same mission. Fortunately for Blacks, both men succeeded. In fact, as early as 1863, a slave named Philip Reed erected and put in place the Statue of Freedom on the Capitol in Washington, D.C. Blacks in the South subsequently outnumbered Whites five to one in skilled trades. And until the passage of Davis-Bacon in 1931, the federal government's policy of accepting the lowest bid on construction projects allowed Black laborers to compete freely for federal work.[7] The Davis-Bacon Act put a sudden end to this democratic capitalism.

In addition to guaranteeing the failure of most Black and Hispanic construction firms, the Davis-Bacon Act keeps young unskilled Blacks and Hispanics from getting laborer jobs and from acquiring building-trade skills and earning a higher wage as a result of the work experience. These are the socially unacceptable and alienated young people who have no other option for work that offers a route to independence. John Cruz, a non-White owner of a construction company in Boston, told the *Washington Times* that Davis-Bacon ensures that the opportunities go to what he calls "privileged workers who make artificially high wages."[8]

The Black unemployment rate in the construction industry in the fourth quarter of 1992 was 26.8 percent, more than twice the rate for Whites. Black- and Hispanic-owned construction businesses are disadvantaged by the legislation because they cannot compete with the taxpayer-subsidized wages for White unionized workers. Small firms owned by ethnic minorities cannot underbid the larger firms. This effectively protects established interests and eliminates competition for government contracts.

Columnist Tony Snow of the *Washington Times* reported the fate of Nora Brazier, owner of a recycling company in Seattle. Ms. Brazier is college-educated and earns $27,000 a year as an accountant. She said because of the Davis-Bacon Act, she is forced "to hire some high school kid and pay him $45,000 to drive a truck. I can't run a business like that."[9]

The Davis-Bacon Act demands that companies such as Ms. Brazier's pay employees the "prevailing wage" for work on federal projects, and the Labor Department uses local union wage scale as a proxy for the prevailing wage, which means paying prohibitive annual union-based salaries of $45,000 for truck drivers, $40,000 for ditch diggers, and $60,000 for carpenters to unskilled workers from low-income groups.

That sort of payroll puts small non-White firms at a severe competitive disadvantage because it doubles their labor costs, which means they cannot underbid large White competitors. Thus, because of the prevailing wage rule, a helper must be paid as a "carpenter" just for hammering a nail.

When $20,000 helpers must be paid as $60,000 carpenters, the cost of building goes up proportionately, as does the rental cost of the finished unit to the low-income-housing residents. The taxpayers pay the difference on all construction projects using more than $2,000 of federal monies—which is one out of every five building projects in the country. That government subsidy adds up to about $2 billion a year from taxpayers. Add in another $100 million a year from small contractors to administer the program. All of that money is spent just to keep Blacks and Hispanics unemployed, which is why the National Association of Minority Contractors joined a group of public housing tenants and Clint Bolick at the Washington-based Institute for Justice in filing a lawsuit challenging the Davis-Bacon Act on race discrimination grounds.

Refusing to challenge Davis-Bacon themselves because they are political opportunists, members of the BUM attacked the motives of Bolick, a White libertarian, on the grounds that he's a conservative and, therefore, automatically a racist. Small wonder the Black rank and file is held down.

While some labor unions block Black self-help projects, no help is forthcoming from the church or civil rights groups. Black politicians in the pockets of the labor movement have never confronted their union benefactors. One of those who remains silent, Representative John Lewis, Democrat from Georgia, has been candid about his loyalties and has characterized PAC money as a "minority entitlement."[10] Lewis, a foe of small business whose U.S. Chamber of Commerce approval rating is zero,[11] boasts accurately that labor's political action committees and a few other PACs made it possible for him to win election in his Atlanta district.

Indeed, during the first six months of 1994, the John Lewis campaign raised $316,568 from PACs. Just four labor-union PACs, from among ninety-four potential lobbies, gave him $205,520, or more than half of his PAC money, according to the Center for Responsive Politics and the National Library on Money and Politics. While Lewis is showered with gratuitous media accolades, calling him "a great civil rights leader," he is amassing an unfair economic advantage over any local challenger for his congressional seat. Yet Lewis feels he has the high moral ground. He has characterized Republican budget cuts as an "onslaught" on children and the poor "reminiscent of crimes committed in Nazi Germany,"[12] as the *New York Times* reported.

While the responses on Davis-Bacon range from silence to approval from the union-influenced civil rights groups and the Democratic Congressional Black Caucus, the National League of Cities has called for its repeal. Former President Bush tried to reform it by creating exemptions for unskilled workers, who would be allowed to work for less than union wages, but President Clinton wasted no time reversing Bush's revisions.

The skilled trade unions acknowledge a resistance to Black members and the practice of "pushing aside" from work assignments those who do belong.[13] Davis-Bacon is another tool.

Vice President Al Gore promised the AFL-CIO that President Clinton would veto any bill that repealed the Davis-Bacon Act. In 1995, when the Senate Labor and Human Resources Committee voted along straight party lines (nine Republicans, seven Democrats) to repeal Davis-Bacon, the Democrats vowed to filibuster the repeal attempt to death. Liberal Democratic Senator Tom Harkin of Iowa promised that union interests would be protected and a repeal of Davis-Bacon "will never see the light of day."[14]

Other than a violation of the Constitution, the destruction of democratic capitalism, and racism, what else is at stake here? The American renewal. How is America helped by a law that keeps low-income Blacks and Hispanics unskilled and unemployed and uncompetitive? How will one more poor Black or Hispanic family help America win back its position as a global leader? How is the country enriched by a law that adds billions to construction costs each year that are paid for by the taxpayers?

That's why my first act as President would be to repeal the federal Davis-Bacon Act and oppose the similar laws in forty states. In doing that, I would be freeing poor people, small firms, and the unions to compete as equals in a free-market economy. I would use the $2 billion saved each year by the repeal of the Davis-Bacon Act to train low-income young people for apprenticeships in the construction industry and for high-tech jobs.

I would also introduce legislation that would prohibit any politician from holding legislative committee assignments or voting on any legislation directly or indirectly related to a special interest from which the politician received a contribution. On Team America, we need all of the available talent, and that is why any form of exclusion, including discrimination against White men, will be viewed as contrary to the national interest.

Davis-Bacon is a classic example of how socialism has fostered racism and harmed the economy. Before 1931, Blacks dominated construction trades because they were the cheapest form of labor and the most skilled, an irresistible combination to an employer seeking the maximum profit. Capitalism is meritorious by its very nature. It is simply bad business to practice racism; it means extra cost, less profit— and national decline.[15]

Capitalism Cures Racism

What is it about capitalism that forces it to be oblivious to race? Competition, writes economist George Reisman in *Capitalism: The Cure for Racism*. If the United States were a capitalistic society, slavery (which is government violence) would never have become an institution and today's racism would be recognized as antithetical to economic growth. As you can see, Davis-Bacon and government intervention in a market economy (even when it disperses $87 billion a year as corporate welfare to private businessess [Aid for Dependent Corporations, or AFDC] or meddles in a baseball strike) prevent the employer and the worker from doing what they respectively consider to be most profitable for them. Democratic capitalism provides workers and employers this choice.

And like slavery, this government coercion is antithetical to capitalism because the employer, worker, and taxpayer are not protected from force (violence in the case of slavery) exercised by a government that in a capitalist economy would protect them from same.

By insisting that government control the means of production and monopolize the ownership of property, anticapitalists socialists and Marxists—as well as proponents of monopoly capitalism—call for more racism and more exploitation of the working classes. The more capitalism—i.e., the greater protection from government intervention—there is, the less racism and class exploitation, because the market will be color blind and the workers' wages will rise naturally as demand for their services rises.

The socialistic intervention not only punishes Blacks with racism, it punishes the taxpayers by forcing them to subsidize the arbitrarily inflated costs. Under democratic capitalism, neither racism nor the exploitation of taxpayers would be possible because competition—low prices, skilled labor—would have rendered race irrelevant. Racism drives up costs and reduces profits. Entrepreneurs are rational and concerned with profits, not perpetuating racism. On the other hand, socialism encourages racism, and it is gradually grinding the American economy into the dust.

Therefore, as the title of Reisman's book states, the cure for racism is capitalism, notwithstanding the Marxist rhetoric. "The freedom of competition under capitalism ensures this result."[16] The sooner Blacks understand that, the more parity they will enjoy. The sooner all taxpayers recognize that racism is a socialistic perversion that undermines our standard of living and demand a smaller and less violent government, the sooner the American dream will become a reality for all of us.

How U.S. Socialism Happened

Underlying all of our nation's economic and social problems is the fact that our country's original system of democratic capitalism has been regulated into a socialist mixed economy since the administrations of Herbert Hoover and Franklin Roosevelt. This bastardized economic system has increasingly stifled our ability to provide skilled jobs, good wages, viable businesses, new homes, and education—all that is necessary to grow the economy. In effect, we have been robbed by the government of our birthright—the freedom to pursue life, liberty, and property. This is no longer possible (it never was for Blacks) because the government fails to protect individual rights, which include privately owned property rights and the right to be protected from force.

Blacks were first shut out of the national economy in the South, where 90 percent of Blacks lived under government-sponsored segregation. By the time Blacks migrated North, capitalism there (where there was free competition for labor) was turned into a mixed economy first by the administrations of Herbert Hoover, and later by the New Deal of Franklin Roosevelt. Thus, Blacks have never lived under capitalism in the United States.[17]

Instead of adhering to the will of the framers of the Constitution and protecting citizens' rights, modern government destroyed democratic capitalism in the South with slavery and other forms of statism. After the Civil War ended in 1865 and before World War I, it began the process that ultimately finished off capitalism in the North with the Hoover and Roosevelt administrations.

Since that period, capitalism has been ground into the dust. We have been burdened with more and more intervention (the force we depend on government to protect us from) and with increasing dependence on government to meet our basic needs for food, clothing, and shelter. Entitlement socialism is simply a manifestation of an economy that is effectively based on authoritarian socialism. The great danger from all of this is not philosophical. There is great potential for violent upheaval when the reality of government debt shatters the illusion of the free lunch. We are now witnessing the early stages of this economic hangover.

Look for discontented working-poor Whites to initiate the beginning of low-intensity military conflicts. The cowardly bombing of a federal building in Oklahoma City may be the opening round from these angry homegrown anarchists to act out their loss of faith in this system. They will likely be joined by entitlement nationalists of all stripes in anti-Republican riots protesting the attempts to cut the growth of government spending. Now that the public is addicted to socialism's entitlements, going cold turkey is out of the question. As a result, the restoration of democratic capitalism will be nearly impossible.

Ironically, if anarchy comes to pass from the left, it will occur under the false pretense of overthrowing a corrupt system of capitalism and its twin companion—racism. Nothing could be further from the truth. Unfortunately, a majority of Blacks and many other Americans equate capitalism with oppression—and oppression with racism. Therefore, because America is (allegedly) a system of capitalism, it is racist, they claim.

In fact, if America were capitalist, it could not be racist. Racism is flourishing because we are awash in socialistic controls.

Black Marxist Hypocrisy

Socialism is a conflict philosophy that believes in a system in which one person can gain only at the expense of another. With this philosophy, socialists have contributed mightily to the prevention of the autarky—the economic self-sufficiency— of the Black community by selling their religion as an economic science that will inevitably arrive to save the world from an exhausted capitalism.

In fact, the United States has demonstrated that liberalism *is* socialism on a slow track and socialism is the early phase of Communism. For example, when forced to distinguish between nationalization and socialization, the Marxian is obliged to admit to the ultimate motive of a state dictatorship.[18]

While Black radical Marxists such as Manning Marable at Columbia University, the vice chairman of the Democratic Socialists of America, and Christian Black Marxist (an oxymoron) Rev. Cornel West at Harvard pump their students and their leftist White comrades with the rhetoric of integration and a workers' utopia, they ignore the failure of the former USSR. Communism—or socialism, as it was formerly called—was doomed to fail because of its insistence on nationalizing private property. Inevitably, as economist Ludwig von Mises predicted in his 1922 book *Socialism*, this would make it impossible to calculate and subsequently plan production. In a socialist state, as we saw in the USSR, murder and oppression were necessary to maintain authoritarian rule.

In a domino effect, the absence of privately owned property leads to no competition from private owners, no market prices, no profit-and-loss system, and, most critically, no consumer-directed producers. The result of these falling economic dominoes is, as writer Bettina Bien Graves explains, that "the planners would not know what to produce, how much to produce, or how to produce it."[19]

It took seventy-two years of shortages, waste, production bottlenecks, mass murders, and political prisoners for the Communist bloc to collapse. In the meantime, American Marxists touted the Communist system and heavily influenced the leadership of Black America. With the exception of many Black American leaders and Marxists, most of the world now understands that Communist planners failed because they could not calculate, and therefore plan, production. They need a market economy to do that.

The absence of planning is also a trait of Black leadership. If you think this criticism is too harsh, the next time you meet a professional Black leader ask for the plan—his, hers, or anyone else's—that involves Blacks in the resolution of their own problems. Your question, I predict, will be followed by uh-uhs, rhetorical flourishes of government responsibility, accusations of White racism, and a final declaration of "fixin' to get ready."

For the Sake of Survival

The only hope for Black America's renewal of excellence is for Blacks to understand the devastating effects of entitlement socialism. The other hurdle is that Blacks must put racism in context. Attacking all Whites as racists, and ostracizing Blacks who speak out about the ineptitude of many Black leaders, serves to brush back criticism. It also preserves a self-serving leadership and ensures the predicament of an out-of-control, twice-exploited underclass. The truth is that White America has been more responsive to solving its historical racial problem and empowering its most volatile minority than any other hegemonic group in the world. But like all numerically dominant groups, it arrogantly suspects that it is dominant because it possesses superior cultural traits.

Cultural baggage notwithstanding, White Americans have enthusiastically, yet naively and ineffectively, supported lavish spending on programs to rectify the damage of legal slavery and three centuries of continuing discrimination and prejudice toward Black Americans. This modern spirit of restitution gradually gave way over the last three decades to skepticism and, in too many cases, to neoracism—a political Southern strategy of both major parties—nativism and academic racism.

White America's restitution investment failed largely because the concept of affirmative action was translated into a middle-class welfare program that benefited upper-crust White women the most. It was structured to make Whites feel virtuous rather than to give disadvantaged Blacks an opportunity to qualify themselves to become competitive.

The Case for Deficit Reduction

As President, I would do all within my power to get Americans working together as a team instead of pitting them against each other. And once I got my Team America together, I would take on this nation's worst enemy—the national debt.

Demagogues thrive on the fact that the voters are short-sighted. Unless we do something quickly, I believe the debt will bury us. Maybe it won't bury you and me, but our children and grandchildren will surely be piled upon. Even those who have no trouble sleeping knowing that we have spent $5 trillion more than we had as a nation might be snapped awake by the fact that the bill is going to have to be paid by their children.

On one of my PBS shows, economist Lester Thurow noted that deficits are "negative savings"—which means that not only are we behind in paying old bills, we are not spending for the future. With a constant budget deficit you have less to invest in research and development, skills, and infrastructure. As an example, Thurow noted that America's global competitors are in a race to see who can build the fastest passenger train. The Japanese have their bullet train at 130 miles an hour; the French version runs at 140; and the Germans have the ICE at 150. What does the United States have? The fastest train in America runs slower now than it did in 1900. Our railroads are critical to this nation's infrastructure investments. It is necessary to maintain our transportation system in order to preserve a competitive edge in moving products back and forth between major centers of commerce.

Thurow said urban transportation systems will be as important in the coming decades as the space race was in the

1960s. "America isn't playing in that game, and if we don't play in that game, we're going to be falling behind the world in terms of industries, and so we're not talking here about just being Puritan and balancing the budget for the sake of balancing it,"[20] he said.

Just as they have neglected our transportation infrastructure, our irresponsible leaders have allowed the ruin of this country's economic infrastructure. It already takes all of the income tax collected from all taxpayers west of the Mississippi to pay just the interest on the debt, which is the fastest-growing part of the budget. Within a few decades, interest expense will exceed 100 percent of the economy, and our economic system will blow up in our faces assuming current trends.

Our government operates in a vicious cycle. More lies. More debt. More poverty. More fear. More racial polarization. More spending. More debt. How can we stop this cycle and still produce a budget with a surplus in order to avoid economic collapse?

A Government That Works

Team America is a communitarian ideal of organized neighborhood self-empowerment associations—people taking care of one another. This idea may seem a bit premature in 1995. But within twenty years, amid massive layoffs among the middle class because of displacement by computers, robots, and cyberspace technology, it will become apparent to more and more people.

The centerpiece of a Tony Brown administration would be my plan to halt the social deterioration caused by the

American mixed economy. Since the American population will never permit a return to Democratic capitalism and a free-market economy, the only politically acceptable solution must be compatible with our socialist-capitalist hybrid form of economic and social organization. Therefore, my proposed framework for welfare reform is Mutual Help Organizations (MHOs), community self-help associations that attack the corrosive effects of our social breakdown and inept government bureaucracies.

The semi-autonomous MHO is modeled on the Japanese *jichikai*. It would develop self-reliant and economically self-sufficient subpopulations within electronically linked neighborhoods. In this twenty-first-century climate, community empowerment means the possession of knowledge and the technical ability to create wealth by distributing information from consumer to consumer in cyberspace. MHOs would be structured to effectively create free-market capitalist incentives and deliver socialist entitlements to their constituencies, while championing moral virtues. Indeed, a citizen group can make a better determination of the social needs of its neighborhood than government bureaucrats in Washington.

The plan calls for the federal government to estabish guidelines and mandate social programs, along with providing the necessary funds to the states and local governments, which, in turn, would subcontract with MHOs to administer and deliver the necessary services to their neighbors.

This self-help scheme will accomplish several objectives. It will return government to the people at the neighborhood level. It will deliver social services more cost-effectively and more humanely. It will prepare an economic and social safety net for those workers who are about to be displaced by computers and machines in the most massive technological unemployment this country has ever known. It will also help people learn to care about and take care of one another. God knows we will need to.

The MHOs can be used to address a great many of our nation's most serious problems, everything from worker displacement by technology to drug addiction to illegal immigration to troubled urban schools and crime. All our nation's problems are compounded by the decreasing buying power of the American people. Families with two providers now earn less in real terms than those with one parent working in the 1960s. The country is crippled by a federal deficit that is expected to rise again from 2.5 percent of GDP in 1997 to 10 to 20 percent of GDP by the year 2030, when it will crowd private borrowers out of the credit market. This expected economic collapse will impact on inner-city Blacks even more severely than other groups, and will lead to more hostility and antisocial behavior among Blacks as well as more White resentment toward Blacks. In a sense, our economic decline has set up Blacks and Whites as mutual scapegoats.

Neighborhood MHOs would offer local solutions, within federal guidelines, that reached into neighborhoods of two hundred to five hundred households. The MHOs are inspired by the Tenth Amendment edict of power "to the people." This method would, in effect, abolish the welfare state and its bureaucracy. It would also avoid transmitting the federal government's codswallop to the local government level.

For example, the MHO in the Woodlawn area on Chicago's Southside should be built around WECAN (Woodlawn East Community and Neighbors). This 15-year-old organization was established to accomplish community empowerment in housing, education, social service, job creation and placement, and economic development. America will be a long way down the road to reforming social services when WECAN's founder Mattie Butler is as well known for saving her community as others are for destroying it.

Government will fund MHOs up to 90 percent. It will give tax incentives that allow donations to MHOs as non-taxable income. It will also eliminate capital gains taxes on individuals and companies on the amount donated to a charity involved in an MHO program. To individuals who invest in MHOs, it would return a retirement benefit—with interest. Businesses that profit from high-technology activities (which cause technological displacement of workers) will pay a 10 percent consumption tax to be earmarked for MHOs. It would be a federal offense for any officer of an MHO to participate in any form of political activity, including voter registration.

MHOs will offer compassionate help from neighbor to neighbor, which transforms people by demonstrating the virtues of personal responsibility. They will inspire compassion and gratitude from those who receive public charity. They will offer training in cyberspace technology for an age in which knowledge is the key resource, diminishing the significance of labor, capital, and raw materials. And MHOs will transform our bureaucratic welfare system into a neighborhood-managed program that temporarily assists people and helps them become productively employed while protecting the children.

The assumption that government can solve social problems is the main reason that social problems have multiplied in tandem with government expenditures. This is another reason for not simply transferring social welfare programs to state governments, which have failed as miserably at welfare-state social engineering as has the government of the United States.

Welfare as We Know It

The government now wants to solve the welfare problem by creating jobs for unwed mothers who have never been in the labor market, are emotionally unstable, or have already been rejected as manual workers because they were deemed educationally unsuitable. To do so, welfare programs will have to expand to accommodate a very expensive support network of child care and an even more costly attempt at training for high-tech jobs. The term "welfare reform" is being used as a covert euphemism for a larger welfare bureaucracy and more government control over family life.

The current bureaucratic approach is also riddled with economic inconsistencies. While the government plans to create new jobs for obsolete workers such as welfare recipients, it is busy fighting inflation by raising interest rates to increase unemployment among the middle class as well. This nation cannot produce enough employment to sustain the living standard of its educated middle class. The more government meddles, the more babies are born into an environment without hope. Our existing welfare disaster is simply perpetuating itself under the duplicitious claim of "ending welfare as we know it," as Arkansas's Willie Clinton likes to put it.

What about "ending entitlements as we know them," as President Brown would propose? Sacrifice and change means big cuts in the sacrosanct Social Security program while guaranteeing all non-wealthy beneficiaries that their payments will not be reduced below their present level. A new, two-tiered plan should be instituted to allow young workers to start profitable retirement plans while the existing Social Security plan is phased out as the older generation passes on.

If entitlement spending (such as Medicare, Medicaid, Social Security, and federal retirement benefits) is combined with the federal debt, it amounts to 60 percent of the federal budget. Unless these programs are reduced, the cost will rise to 70 percent of federal spending by the year 2003. Within thirty years, the major entitlements will probably consume 100 percent of federal income.[21] The result: anarchy and no entitlements for anyone.

Cleaning Up the Books

Can you imagine living in an America where you can decide how to spend the money you earn—an America without an income tax? First and foremost, a President Tony Brown would make that dream a reality. I pledge to you that my administration would not only eliminate the federal tax on personal income, but because of the zeroing out of the corporate profit tax and the capital gains tax, America would attract unheard-of amounts of investment capital into what is now a declining power. We could then produce more jobs than burgers for the first time in history. You can sum up my administration in two words: growth and prosperity. A Tony Brown administration also means a balanced budget— not just reduced annual deficits—and no accumulated debt. This alone would lower interest rates for home loans and businesses, which means more jobs. Modeled after a plan by economist Thurow, my proposal is designed to balance the federal budget, eliminate the $200 billion interest payment, and create sufficient capital for research and development. In addition, it will reward productivity, increase saving, and act as a brake on consumption.

Under my plan, the essentially unfair personal income tax would be replaced by a non-regressive flat-tax-rate national retail sales tax—exempting real estate and securities. I do not favor a flat tax because it is a tax on income.

A national sales tax is simple, fair, and progrowth. The counterproductive personal-income, corporate (a hidden consumption tax that hits the poor hardest), and capital gains taxes would be eliminated.

On the subject of helping the unemployed, it takes a little sophistication, but not a lot of brains, to realize that an adamant hatred of people fortunate enough to have assets, and class warfare do not create jobs. Whether a lower capital gains rate benefits the wealthy more than the middle class is academic, because the top 3 percent of taxpayers pay over 40 percent of all taxes and over 50 percent of capital gains go to people who earn less than $50,000. A lower rate on the capital gains tax owed on property like stocks will unleash more investment capital, which in turn will create more businesses and new jobs for poor low-income people.

The capital gains tax could be retained if earmarked exclusively for MHOs. The payroll tax would be reduced by one-third. These revenue losses would be offset by a national retail sales tax and a value-added tax for businesses. I would also index cost-of-living allowances (COLAs) for entitlement benefits to the GDP—not the rate of inflation. Thus the beneficiaries would not be guaranteed increases when the economy was sinking everyone else with inflation.

My plan would allow taxpayers to increase the nation's saving rate and at the same time control the size of the government, which would put a bridle on tax-and-spend Democrats and on borrow-and-spend Republicans. Under this system, term limitations of elected officials would not be necessary, because each taxpayer could indirectly control government revenues and the size of government through his or her spending or saving habits. Saving would be involuntary and capital finance could be freely invested to spur growth.

Another major boost to the federal Treasury would result from a national sales tax. Presently, for every $1 of tax not paid on illegal income, $9 of taxes are not paid on legal income. Unreported legal income is estimated to be in excess of $150 billion.[22] Therefore, another $150 billion a year would be collected in taxes from the underground economy—everything from antiques dealers to drug dealers[23] who don't report their incomes.

Taxes lost through unreported income would be automatically and efficiently collected at the point of purchase through consumption and value-added taxes. For example, when the untaxed drug pusher bought a $100,000 car, he or she would pay federal and local taxes during the transaction and the retailer would be compensated for collecting them. A national sales tax, according to a Cato Institute study, will double the savings rate and increase the average American's standard of living anywhere from 7 to 14 percent.[24]

Simply reforming the personal income tax alone would not raise the necessary revenues, and such taxes are inevitably unfair to some people, but taken together, a 10 percent value-added tax on business at each stage of production and a 15 percent retail sales tax made non-regressive by exempting the first $5,000 of purchases for each person would create a tax code based on fairness. In fact, it allows a family of four with an income of $20,000 to pay no taxes at all.

The issue is not avoiding new taxes, as Republicans

claim, or more taxes, as the Democrats preach; it is tax fair-
ness. The most imminent danger is posed by the fact that
the country's economy is in the terminal stage of a thirty-
year bipartisan deficit binge. The nation simply cannot sur-
vive many more years of the same fiscal and financial abuse.

As the following chart illustrates, instead of a deficit of
$365 billion in 1992, the surplus would have been $132 bil-
lion under my proposal—a surplus equal to 1 percent of the
GDP. And, of course, there is the $2 billion bonus that could
be saved each year from the repeal of the Davis-Bacon Act.
This plan, including the elimination of several cabinet-level
departments, could eliminate the deficit immediately and
abolish the national debt within four years—which saves an-
other $300 billion a year in interest payments.

TONY BROWN ADMINISTRATION PROPOSED TAXATION

Federal Revenue 1992 (in Billions of Dollars)

	EXISTING LAW	PROPOSED LAW
Personal income tax	$479	$0
Retail sales tax (15%)	0	473
Corporate profit tax (a stealth sales tax)	89	0
Indirect business tax	97	97
Social insurance tax	411	292
Value-added tax (10%)	0	537
Saddam Hussein Gas Tax ($1.50 per gallon)	0	174
Total	$1,076	$1,573
Expenditures	$1,441	$1,441
Surplus or deficit	−$365	+$132

Department of Growth

As President, I would put together a team of experienced and realistic individuals to focus on economic growth and the development of human resources. By implementing a new surplus-producing tax code, I would lead an assault against government size, debt, and out-of-control entitlements. I would ask the American public for sacrifice and change and explain why it is necessary for our nation's survival.

I would ask self-described "bleeding-heart conservative" Jack Kemp, a visionary public servant, to supervise a *Department of Growth* as Vice President or Secretary of the Treasury. And directly under him would be the management of the Mutual Help Organizations.

Facing Up to the Truth

There would be no avoidance of the truth in my administration. Lies will only guarantee more deficits, sociopathic leadership, and societal breakdown. For example, one truth that Americans need to face is whether to pay more—a lot more—for gasoline now, or to spend billions more in dollars and human lives in a war that will inevitably result from our dependence on the Middle East for oil.

I would impose a Saddam Hussein Gas Tax of $1.50 per gallon, because at $1 a gallon for gasoline, the American public can afford to continue its dangerous dependence on overseas oil while neglecting the development of alternatives such as the electric car and public transportation. But with the Saddam Hussein Gas Tax, this country will be forced to face the reality that down the road, it will either have to go to war to seize the oil reserves of other nations or find alternative modes of energy and transportation. Financially and morally, the United States simply cannot afford to fight oil wars anymore. There are too many costly battles at home that need our attention and our solutions.

Here are summaries of other major areas of focus in the Tony Brown Presidency:

Drugs and Crime. To eliminate the drug problem, I would eliminate demand. I would propose legislation to target drug buyers for arrest. They would be given the choice of standing trial and facing imprisonment or placement in a drug treatment program. Chronic drug users and addicts would be targeted also for therapy and treatment programs in order to eliminate demand for drugs in this country.

Under my crime policy, rehabilitation would be the focus for first-time, nonviolent offenders. But repeat and violent offenders would be severely punished. First-time nonviolent offenders would be placed in boot camps that reward good behavior with training in high-tech, marketable skills designed to cut recidivism. Repeat and violent offenders would do hard time on fixed sentences in maximum-security prisons with no recreational facilities, no opportunities for social interaction, no conjugal visits, no parole, and zero tolerance of gang activity. Through a demerit system, individual members of parole boards and all criminal court judges would be held accountable for future crimes committed by parolees or probationers. Excessive demerits would result in removal of judges from the bench.

Education. I would propose a school voucher system in which parents would be allowed to purchase an education for their children from the best provider available to them, whether or public or private. This is the informal system that is already widely in practice but only available to the well-off. The Clintons send their daughter to a private school, as do 50 percent of urban public school teachers with school-age children. Perhaps that is why 85 percent of Blacks and Hispanics prefer a voucher system themselves, according to a Gallup poll.[25] Merit pay and sizable cash bonuses would also be provided to reward public school teachers and administrators for excellence in performance by their students. My education package would also include subsidies for career training and youth high-tech training programs modeled after the German apprentice system.

Human Rights. My administration would deal with international human rights violations. In particular, I would move to expose a twenty-first-century form of slavery that Americans have heard virtually nothing about. Today in northern Africa, Sudanese children are being sold into slavery for $15 each. According to the American Anti-Slavery Group, Blacks are sold to Arabs in North Africa, Mauritania, and Sudan as "chattel: used for labor, sex and breeding." In Sudan, Arab militia raid Dinka tribal villages and kill the men in order to enslave the women and children. The Qur'ān forbids this heinous practice, but today's informal practice continues unabated. Rape is commonplace, and the slaves' Achilles tendons are cut, crippling them for life.[26] The Congressional Black Caucus is slow to understand the issue, which does not surprise me.

Racial Equality. As President, I would initiate legislation to issue a formal apology to the American descendants of

Africans for the kidnapping of their ancestors in their home-
land, the agony suffered as a result of government-sanctioned
slavery, and the long-term deleterious effects on these
shores. It has been one of the great atrocities of civilization
and a blot on the generally virtuous American character.
Thirty-two million Americanized Africans deserve no less
than the respect our government has shown Native Ameri-
cans and the Japanese-Americans who were forced from their
homes and land.

I also would encourage those Americans who do not
want pluralism and inclusion to respect the laws of the land
that equally protect the rights of other citizens. For instance,
I would encourage and legally support the choice of some
Blacks who want to separate from Whites and cluster within
the United States borders or migrate to another country. It
is my opinion that people who make this choice will be
immensely more productive in communities that are cultur-
ally more amenable, and their departure will enhance those
communities they leave behind. An example not of racial
but religious preference is those Americanized-African Jews
who have migrated to Israel.

I would specifically encourage disaffected Blacks to mi-
grate to sparsely populated states—that's what the White su-
premacy groups are already doing—and politically control
them. This will replace Congressional gerrymandering to
create Black-minority districts. This will give the Blacks
who choose to live in a Black-controlled state the chance to
learn governance and free-market economics. It will also
teach them that neither nationalism nor noneconomic so-
cialism feeds hungry people, picks up their garbage, or pro-
vides jobs. Let Black socialists set their school curricula. If
students can become more useful reciting slogans than work-

ing at capitalist jobs, let them. Encourage them to run a state and educate, protect, and serve the people.

This may sound like a revolutionary idea, but in fact it is already practiced by the Mormons, who have congregated in Idaho and Utah and control those states politically and economically. Interestingly, there are Blacks in those states, and the 11,576 Black Utahans are among some of the most influential residents. The crime and unemployment rates among Utah's Blacks are very low, and their standard of living is above average. This demonstrates that an ethnocentric dominant group can create an environment in which a numerical minority can prosper.

There are over three million Blacks in New York City alone and a total of 32 million in all of the United States. If only three million of that total moved to a state or a cluster of designated states, the separation and empowerment various groups such as the Nation of Islam preach would be a reality. Any separatist leader could easily become the governor of Montana, South Dakota, Delaware, North Dakota, Vermont, Alaska, or Wyoming if only 800,000 Blacks would follow him or her to any of those states and vote. There would also be two Black U.S. senators, a National Guard, a Congressional delegation, and matching grants they could fight over from each state they dominated. Of course, these governments would have to pick up the garbage and protect the rights of the White minority.

Ending Welfare

To understand the magnitude of the welfare problem, take the case reported by the Boston Globe of an unmarried woman from Puerto Rico who moved to Boston in 1968, went on welfare, and moved into public housing. By 1995, she had seventeen children, seventy-four grandchildren, and fifteen great grandchildren. The fourth generation is now on welfare, and, according to David Brinkley on ABC-TV, "few, if any of them, have ever worked."[27] The cost to taxpayers is reported to be $1 million a year for this family alone.

As you can see from this admittedly extreme example, the fruits of government welfare policy are: unemployment and poverty. Or its more subtle by-products: undereducation and computer illiteracy, a modern form of miseducation. The economic and social problems facing the nation and especially our inner cities and hardscrabble rural ares will not and cannot be solved by the standard government approaches of educating only college-bound youth for the obsolete workforce of 1963 and socially promoting the lowest-income stratum through school so they can receive a diploma that some of them can't even read.

This explains why their failure is fairly predictable. Only unemployment, poverty, and criminality can follow this system of miseducation and technical illiteracy in a global enonomy. And these factors produce inadequate mothers and fathers, who, in turn, guarantee more broken homes.

The federal bureaucracy then exacerbates its own failed invention by repeating the cycle with welfare programs that are guided by the principle that the problem is not government but unwed mothers and their children. The government doesn't stop there. The socialist planners see no contradiction in government-subsidized female-headed homes that have created and perpetuate a matriarchal welfare culture that empowers women and subordinates men and boys. Without the provider role, males in this culture are stripped of their psychological worth. These men are forced out of the traditional role that we culturally demand from American males. Case in point: After racism robbed many disadvantaged Black men of their jobs, the government stole their families. Combined, these two forces are responsible for the disintegration of manhood of many of the men who grew up in this government-sponsored dysfunctional matriarchy.

I propose the following guidelines for the MHOs to administer welfare programs at the neighborhood level. Recipients would have to accept training in apprentice programs teaching computer skills and other specific employment skills. Payments to women on Aid to Families with Dependent Children would be cut or dropped after the second baby or after a specified number of years on the dole. Alternative services for the dependent children, such as food stamps and health coverage under Medicaid, would be provided. The savings would go to retraining mothers who want to free themselves of welfare dependency.

Facing Reality

Many White voters are angry at Black people because they blame them for spiraling welfare costs, violent crime, and chronic urban problems. Many Blacks are convinced that there is a White cultural conspiracy aimed at keeping them down.

Blacks and Whites are equally wrong about each other. It seems to me that a more reasonable explanation may be found by examining the opinion polls. They show that a majority of Americans believe that the country is going in the wrong direction.

Black and White racism breeds a new American virus of hatred and threatens every aspect of American life. The Black predicament is at the core of America's crisis of crime, debt, drugs, broken families, and lost economic competitiveness. The two major parties have failed to see the rehabilitation of Black people as the key to settling the American dilemma.

The issue of Black America has been tabled even by the Democrats, whose power base is the Black vote. For the first time since 1944, the rehabilitation of the Black community was not even discussed in the last presidential race, out of the fear that it would offend White voters in the Democratic primaries. Perhaps it would have. But voters cannot always be told just what they want to hear. Only the truth will turn this nation around.

We cannot ignore the national debt that is dragging down our standard of living and threatening to make welfare recipients of our children. We cannot ignore the school systems, whose biggest problem is not those who fail to graduate, but those who do and can neither get into college nor find a job because they are functionally illiterate. We must not ignore the lunatics on the fringes of both the Black and White communities who are doing everything in their power to draw the mainstream into a race war that would destroy democracy for everyone. And we can no longer afford to tolerate sociopathic leaders in the White House and Congress.

We are in a national state of denial. The cure is to get in touch with reality. Certainly, America is not going to survive as a superpower by ignoring the Black predicament.

When Clinton's term expires, we will be burdened by a national debt grown by $1.3 trillion.

The decline of moral virtues in our national character is costly. Our major institutions of law enforcement, politics, church, media, and business have become special interest lobbies, not instruments for the common good. Under our current legal system, "O.J." justice is for sale. The United States is facing moral defeat.

Saving America

Without a halt to the moral decline through a renewal of moral virtues in a new American character, the elimination of the debt, and the reinvention of the Black community as a competitive force in a free-market economy, the United States is probably headed for the status of a Third World nation. Or, perhaps, our nation faces a future as an authoritarian socialist state in which the weakest population segments will be allowed simply to wither and die through an official government policy of triage.

As I've noted, the symptoms of societal disorganization that become apparent first in the most marginal segments— the underclass—will eventually afflict the middle and upper classes. It follows, then, that to save White America, we must rebuild Black America—and vice-versa. You and I may not like one another, but we're all we've got.

The national debt and excessive spending are destroying the infrastructure of the economy, and the most marginal groups get hit first. Young White children, as well as Black, Brown, Yellow, and Red children, are now at risk in the United States. Although the fact is underreported by the media, half of the nation's poor are Whites, and the rate of poverty among White children in rural areas is higher than it is among Black and Hispanic children in three of the top five urban areas. In fact, the poverty rate for Whites in America, according to the Center on Budget and Policy Priorities, is higher than it is in France, Germany, or Great Britain.

The American renewal that I envision is centered on personal and social ethics as well as a competitive and growing economy. America is decaying morally, and the symptoms are evident in our material environment as well as in our government-without-ethics. A society without moral virtues is a restless and uncertain society. The inner cities, for instance, reflect a gradually encroaching societal instability. I believe we can begin to rebuild the inner cities of America by beginning at the top and rebuilding the moral and ethical fiber of this country's defining institutions.

The West created an arsenal of lethal weaponry to destroy Communism, and one day we woke up and the Berlin Wall had come down, and Communism along with it, without a single shot. None of our tanks or guns did us any good, because something else was going on. The human spirit was being unleashed—something much mightier than firepower. The Germans changed. And when they changed internally— when they experienced a mental metamorphosis, a spiritual evolution—that changed their external reality.

America must face this inner challenge the same way the Germans faced their challenge of unification. If Americans do not learn to change and sacrifice, Americans will not grow as a people. In Germany, the rich capitalist West is absorbing the poor socialist East as they exponentially evolve into a higher social and economic order. In

America, the underclass presents this challenge to the upper classes; Blacks present it to Whites; and Whites present it to Blacks.

If we do not adapt as one people, as Team America, we will not survive as a superpower. We have a choice—to grow as a people or to dissolve as a republic.

NOTES AND SOURCES

Chapter 1: Different Ship, Same Boat

1. David S. Broder, "Illegitimacy: An Unprecedented Castastrope," *The Washington Post*, June 22, 1994, p. A21.
2. Ibid.
3. Mickey Kaus, "Sighting a Great White Threat," *Newsweek*, May 30, 1994, p. 48.
4. Charles Murray, "The Coming White Underclass," *The Wall Street Journal*, October 29, 1993, p. A14.
5. Kaus, "Great White Threat," p. 48.
6. Marc Cooper, "Reality Check," *Spin*, July 1994, p. 52.
7. Ibid., p. 53.
8. Ibid., p. 52.
9. Ibid., p. 54.
10. J. Freedman, *From Cradle to Grave* (New York: Simon & Schuster, 1993), p. 58.
11. Lester Thurow, *Zero-Sum Solution* (New York: Simon & Schuster, 1985), p. 65.
12. Ibid.
13. "One on One with Lester Thurow, Part I," *Tony Brown's Journal* (1514), PBS, May 1, 1992, and "One on One with Lester Thurow, Part II," *Tony Brown's Journal* (1515), PBS, May 8, 1992.
14. Luke 16:18–21.
15. Luke 16:24.
16. Luke 16:29–32.

17. Sam Roberts, "The Greening of the Black Middle Class," *The New York Times*, June 18, 1995, p. E1.
18. Ibid., p. 4.
19. Kimberly J. McLarin, "A Child Shines Amid the Shambles," *The Philadelphia Inquirer*, April 14, 1992, p. A1.
20. Ibid.

Chapter 2: Entitlement Socialism: Why America Is Not Working for U.S.

1. Peter Peterson, *Facing Up: Paying Our Nation's Debt and Saving Our Children's Future* (New York: Simon & Schuster, 1994), p. 100.
2. David Kline, *Critical Intelligence*, April 1994, p. 12.
3. Ibid.
4. Michael C. Dawson, "Black Discontent: The Preliminary Report on the 1993–1994 National Politics Study," University of Chicago, 1994, p. 7.
5. Jonthan Rauch, "The Hyperpluralism Trap," *The New Republic*, June 6, 1994, p. 22.
6. Richard Lacayo et al., "Remember Deficit," *Time*, November 8, 1993, p. 39
7. "Is America Going Bankrupt?" *Tony Brown's Journal* (1606), PBS, March 5, 1993.
8. Peter Peterson, "Entitlement Reform," *The New York Times Book Review*, April 7, 1994, p. 39.
9. Jonathan Karl, "Taxpayers of Future Face 82% Wallop," *New York Post*, February 9, 1994. p. 2.
10. Ibid.
11. Peterson, "Entitlement Reform."

12. Bob Woodward, *The Agenda: Inside the Clinton White House* (New York: Simon & Schuster, 1994), p. 214.
13. Ibid.
14. Lester Thurow, speech, Milton S. Eisenhower Symposium, Johns Hopkins University, Baltimore, MD, October 18, 1988.
15. Alfred L. Malabre, Jr., *Beyond Our Means: How America's Long Years of Debt, Deficits and Reckless Borrowing Now Threaten to Overwhelm Us* (New York: Vintage Books, 1987), p. 49.
16. Peterson, *Facing Up*, p. 100.

Chapter 3: The Failure of Black America's Leaders

1. Anemona Hartocollis, "Walking His Own Line," *New York Newsday*, April 14, 1993, p. 60.
2. Barbara Reynolds, "Blacks Should Use Their Billions to Call Their Own Shots," *USA Today*, July 31, 1992, p. A9.
3. Claude McKay, *Harlem: Negro Metropolis* (New York: E. P. Dutton, 1940), p. 262.
4. Ibid., p. 184.
5. DeWayne Wickham, "Affirmative Action: Case of Friendly Fire," *USA Today*, March 20, 1995, p. A13.
6. Manning Marable, *How Capitalism Underdeveloped Black America* (Boston: South End Press, 1983), p. 138.
7. Manning Marable, "The Divided State of Black Leadership," *U.S. News & World Report*, July 18, 1994, p. 29.
8. Joey Merrill, "Mr. Chavis Goes to Washington: Rein-

venting the NAACP," *Diversity & Division: A Critical Journal of Race and Culture* (Fall 1993):21.

9. Leon Weiseltier, "The Decline of the Black Intellectual," *The New Republic*, March 6, 1995, p. 34.

10. Cornel West, "The Dilemma of the Black Intellectual," *The Journal of Blacks in Higher Education* (Winter 1993/1994):64.

11. Ibid., p. 61.

12. Ibid., p. 64.

13. Harold Cruse, *Plural but Equal: A Critical Study of Blacks and Minorities in America's Plural Society* (New York: William Morrow/Quill, 1987), p. 75.

Chapter 4: Conspiracies and Black America

1. Jason DeParle, "Talk of Government Being Out to Get Blacks Falls on More Attentive Ears," *The New York Times*, October 29, 1990, P. B7.

2. Stephen G. Thompkins, "In 1917, Spy Target Was Black America," *The Commercial Appeal* (Memphis), March 21, 1993, p. A7.

3. Ibid., p. A1.

4. Major J. E. Spingarn, "Advisory Committee to the Chief of Staff of the Army," Memorandum to Col. Churchill, Subject: "Negro Subversion," June 10, 1918.

5. Ibid.

6. David Levering Lewis, *W.E.B. DuBois: Biography of a Race* (New York: Henry Holt and Company, 1993), p. 555.

7. Ibid., p. 556.

8. Ibid., p. 468.

9. Stephen G. Thompkins, "Top Spy Feared Current Below Surface Unrest," *The Commercial Appeal* (Memphis), March 21, 1993, p. A8.

10. Stephen G. Thompkins, "Army Feared King, Secretly Watched Him," *The Commercial Appeal*, (Memphis), March, 21, 1993, p. A7.

11. Ibid., p. A8.

12. Ibid.

13. Ibid., p. A9.

14. Ibid., p. A8

15. Kenneth O'Reilly, *Racial Matters: The FBI's Secret File on Black America, 1960–1972* (New York: The Free Press, 1989), p. 275.

16. Ibid., p. 122.

17. David J. Garrow, *Bearing the Cross: Martin Luther King, Jr., and the Southern Christian Leadership Conference* (New York: William Morrow, 1986), p. 188.

18. O'Reilly, *Racial Matters*, p. 269.

19. Ibid., p. 268.

20. Garrow, *Bearing the Cross*, p. 555.

21. Ibid.

22. O'Reilly, *Racial Matters*, p. 152.

23. Ibid., p. 153.

24. Carl Rowan, "Martin Luther King's Tragic Decision," *Reader's Digest*, September 1967, pp. 37–42.

25. Ibid., p. 42.

26. Carl Rowan, *Breaking Barriers: A Memoir* (New York: Little Brown & Company, 1991), p. 293.

27. Ibid.

28. Ibid, p. 305

29. Garrow, *Bearing the Cross*, p. 577.

30. Kopin Tan, "Jack Ruby Letter Claimed LBJ Role in JFK Slay Plot," *New York Newsday*, April 15, 1995, p. A6.

31. Stephen G. Thompkins, "Spying Linked Carmichael to Chinese, Cuba," *The Commercial Appeal* (Memphis), March 21, 1993, p. A9.

32. Ibid., p. A7.

33. Ibid.

34. Ibid., p. A9.

35. Ibid.

36. Ibid., p. A7.

37. Juan Williams, "Roads from the Riots," *The Washington Post*, April 3, 1988, p. B1.

38. David Levering Lewis, "The Intellectual Luminaries of the Harlem Renaissance," *Journal of Blacks in Higher Education* (April 1995):68.

39. Tony Martin, *Race First: The Ideological and Organizational Struggles of Marcus Garvey and the Universal Negro Improvement Association*, The New Marcus Garvey Library, no. 8 (Dover, MA: The Majority Press, 1976), p. 274.

40. David Levering Lewis, "Parallels and Divergencies: Assimilationist Strategies of Afro-American and Jewish Elites from 1910 to the Early 1930s," *The Journal of American History* 71, no. 3 (December 1984):560.

41. Ibid., p. 562.

42. Tony Martin, *Literary Garveyism: Garvey, Black Arts and the Harlem Renaissance*, The New Marcus Garvey Library, no. 1 (Dover, MA: The Majority Press, 1983), p. 159.

43. Lewis, "Parallels and Divergencies," pp. 563–564.

44. Ibid.

45. Ibid.

Chapter 5: Fear of Genocide

1. "The 'Bell Curve' Agenda," *The New York Times*, October 24, 1994, p. A16.
2. Richard Herrnstein and Charles Murray, *The Bell Curve: Intelligence and Class Structure in American Life* (New York: The Free Press, 1994), pp. 525–526.
3. Leonard C. Lewin, *Report from Iron Mountain on the Possibility and Desirability of Peace* (New York: The Dial Press, 1967).
4. Leonard C. Lewin, *Triage* (New York: The Dial Press, 1972).
5. "Finds Touch of Africa in 28 Million Whites," Associated Press, June 15, 1958, p. 72.
6. Herrnstein and Murray, *The Bell Curve*, pp. 525–526.
7. Ibid.
8. Ibid.
9. Malcolm W. Browne, "What Is Intelligence, and Who Has It?" *The New York Times Book Review*, October 16, 1994.
10. "The 'Bell Curve' Agenda," p. A16.
11. E. J. Dionne, Jr., "Race and IQ: Stale Notions," *The Washington Post*, October 18, 1994, p. A17.
12. Patricia A. Turner, *I Heard It Through the Grapevine: Rumor in African-American Culture* (Berkeley: University of California Press, 1993), p. 183.
13. L. Fletcher Prouty, *The Secret Team: The CIA and Its Allies in Control of the United States and the World*, (Costa Mesa, CA: Institute for Historical Review, 1973), p. viii.
14. Turner, 2nd ed., p. 184.
15. Ibid., p. xvi.

16. Barbara Harff, "The Etiology of Genocides," in *Genocide and the Modern Age: Etiology and Case Studies of Mass Death*, ed. Isidor Wallimann and Michael N. Kobkowski (New York: Greenwood Press, 1987), p. 46.

17. John K. Roth, "Genocide, the Holocaust, and Triage," in *Genocide and the Modern Age: Etiology and Case Studies of Mass Death*, ed. Isidor Wallimann and Michael N. Kobkowski (New York: Greenwood Press, 1987), p. 83.

18. Rodney Stark and James McEvoy III, "Middle Class Violence," *Psychology Today*, November 1970, p. 111.

19. Ibid.

20. Ibid.

21. Ibid.

22. Harff, "Etiology of Genocides," p. 41.

23. John Chester Miller, *The Wolf by the Ears: Thomas Jefferson and Slavery* (New York: The Free Press, 1977), p. 38.

24. Ibid.

25. Ibid., p. 41.

26. Ibid.

27. Ibid.

28. Ibid., pp. 154–155.

29. *A Few Good Men*, Castle Rock Entertainment, Columbia Pictures release, 1993.

30. Virginia Breen and Jere Hestur, "Khalid Blasts Black Pols," New York *Daily News*, March 30, 1994, p. 26.

31. Lynne Duke, "Blacks, Asians, Latinos Cite Prejudice by Whites for Limited Opportunity," *The Washington Post*, March 3, 1994, p. A9.

32. Harff, "Etiology of Genocides," p. 57.

33. "Take No Prisoner," *Daily Challenge* (New York), March 31–April 3, 1994, p. 3.

Chapter 6: Negative Images, Negative Results

1. Marc Elrich, "Stereotype Within," *The Washington Post*, February 13, 1994, p. C4.
2. Ibid.
3. Carter G. Woodson, *The Miseducation of the Negro* (1933; reprint, Trenton, NJ: Africa World Press, Inc., 1990), p. xiii.
4. Robert E. Cole and Donald R. Deskins, Jr., "Errata," *California Management Review* (Spring 1989):160.
5. Robert E. Cole and Donald R. Deskins, Jr., "Racial Factors in Site Location and Employment Patterns of Japanese Auto Firms in America," *California Management Review* (Fall 1988):17.
6. Ibid., p. 18.
7. Ibid., pp. 17–18.
8. "The Japanese Overseas Businessman in Japanese Literature," paper presented at the East Asian Investment in Arizona's Future Conference, Arizona State University, Tucson, February 19–20, 1988, in Cole and Deskins, "Racial Factors," p. 18.
9. Cole and Deskins, "Racial Factors," pp. 14–15.
10. Woodson, *Miseducation*, p. xiii.
11. Jack Newfield, "Jesse Jackson's Mid-Life Crisis," *Penthouse*, August 1994, p. 118.
12. Leon Wieseltier, "All and Nothing at All: The Unreal World of Cornel West," *The New Republic*, March 6, 1995, p. 35.
13. U.S. Census Bureau, *1990 Census* (Washington, D.C.: U.S. Department of Commerce, 1990).
14. Ibid.

15. Charisse Jones, "Years on the Integration Road: New Views of an Old Goal," *The New York Times*, April 10, 1994, p. A1.
16. Martin Bernal, *Black Athena*, vol. 1 (New Brunswick, NJ: Rutgers University Press, 1987), p. 82.
17. Ibid., p. 79.
18. Ibid., p. 330.
19. Sam Roberts, "In Middle-Class Queens, Blacks Pass Whites in Household Income," *The New York Times*, June 6, 1994, p. 1, and Sam Roberts, "The Greening of the Black Middle Class," *The New York Times*, June 18, 1995, p. E1.
20. Barbara Reynolds, "Black Teens Are Leaders in Saying No to Drugs," *USA Today*, March 4, 1994, p. 9A.
21. Theodore Cross, "Black Africans Now the Most Highly Educated Group in British Society," *The Journal of Blacks in Higher Education* (Spring 1994):92.
22. Martha Bayles, *Hole in Our Soul* (New York: Scribner's, 1994), p. 351
23. Nelson George, *Buppies, Bboys, Baps & Bohos* (New York: HarperCollins, 1993), p. 92.
24. Chuck Philips, "Rap Finds a Supporter in Rep. Maxine Waters," *Los Angeles Times*, February 15, 1994, p. F1.
25. Ibid.

Chapter 7: "AIDS" or "DAIDS"?

1. "The AIDS Cover-up," *Tony Brown's Journal* (1534), PBS, November 27, 1992.
2. Peter Duesberg and Bryan Ellison, *Inventing the AIDS Virus* (Washington, D.C.: Regnery Gateway, 1995).
3. Ibid., p. 223.
4. "AIDS Cover-up."

5. Joyce Mullins, "Conference Speaker Differs with Many on Cause of AIDS," *Delaware State News*, October 13, 1993, p. 1.

6. Ibid.

7. Gerald B. Dermer, *The Immortal Cell* (Garden City Park, NY: Avery Publishing Group, 1994).

8. D. S. Greenberg, "Health-Care Spending—Up Up, and Away," *Lancet* (1992):1086–1088.

9. "Hunting the Virus Hunter," *Tony Brown's Journal* (1410), PBS, May 10, 1991.

10. *HIV/AIDS Surveillance Report*, Centers for Disease Control and Prevention, Atlanta, December 1993, p. 8.

11. "Sex American Style," *The New York Times*, October 8, 1994, p. 22.

12. Robert E. Willner, *Deadly Deception* (n.p., Peltec Publishing Co., 1994), p. 20.

13. Gina Kolata, "New Picture of Who Will Get AIDS Is Dominated by Drug Addicts," *The New York Times*, February 28, 1995, p. C3.

14. "Hunting the Virus Hunter."

15. John Lauritsen and Hank Wilson, *Death Rush: Poppers and AIDS* (New York: Pagan Press, 1986), p. 60.

16. "A Death Sentence?" *Tony Brown's Journal* (1805), PBS, March 17, 1995.

17. Laurie Garrett, *The Coming Plague* (New York: Farrar, Straus & Giroux, 1994), p. 356.

18. Peter Duesberg and John Yiamouyiannis, *AIDS* (Delaware, OH: Health Action Press, 1995), p. 7.

19. Tom Bethell, "T-Cells and C Notes," *The American Spectator* (April 1995):16.

20. Christine Johnson, "The Role of HIV in AIDS," *PRAXIS* (Summer/Fall 1994):26.

21. Ibid., p. 29.

22. Ibid.
23. "The Cult of the Condom," *New York Post*, July 31, 1992, p. 20.
24. Pearce Wright, "Smallpox Vaccine Triggered AIDS Virus," *Times* (London), May 11, 1987, p.l.
25. Alexander Langmuir, "Biological Warfare Defense," *American Journal of Public Health* 42 (1952):235–258.
26. Duesberg and Yiamouyiannis, op. cit., p. 102.
27. "Hunting the Virus Hunter."
28. Duesberg and Yiamouyiannis, op. cit., p. 90.
29. "Hunting the Virus Hunter."
30. Ibid.
31. *Lancet*, April 1993.
32. "AZT," *Day One*, ABC-News, October 18, 1993.
33. Celia Farber, "AIDS: Words from the Front," *SPIN*, April 1994, p. 83.
34. "AZT."
35. Ibid.
36. Ibid.
37. Ibid.
38. Celia Farber, "Fatal Distraction," *SPIN*, June 1992, p. 44.
39. Garrett, *Coming Plague*, p. 383.
40. Patrick Pacheco, "The Pos 50," *Poz*, August/September 1994, p. 67.
41. "Hunting the Virus Hunter."
42. Ibid.
43. Duesberg and Yiamouyiannis, op. cit., p. 97.
44. "AIDS: The Syndrome in a Nutshell," *HEAL*, Spring/Summer 1994, p. 7.
45. "Hunting the Virus Hunter."
46. Ibid.
47. Ibid.

48. Ibid.

49. Jad Adams, *AIDS: The HIV Myth* (New York: St. Martin's Press, 1989), p. 129.

50. Garrett, *Coming Plague*, p. 363.

51. Ibid., p. 364.

52. Adams, *AIDS*, pp. 191–192.

53. Ibid., p. 199.

54. Thomas A. Bass, *Camping with the Prince and Other Tales of Science in Africa* (New York: Houghton Mifflin, 1990), p. 243.

55. Ibid., p. 53.

56. Ibid., p. 43.

57. Ibid., pp. 247–248.

58. Merck & Co., *Merck's Manual* (Rahway, NJ: Merck & Co., 1992).

59. Robert Root-Bernstein, *Rethinking AIDS: The Tragic Cost of Premature Consensus* (New York: The Free Press, 1993), p. 45.

60. "The Way Things Aren't: Rush Limbaugh Debates Reality," *EXTRA!* Fairness & Accuracy in Reporting, July/August 1994, p. 15.

61. Ibid.

62. Ronald F. Carey et al., "Effectiveness of Latex Condoms as a Barrier to Human Immunodeficiency Virus-Sized Particles Under Conditions of Simulated Use," *Sexually Transmitted Diseases* 19, no. 4. (July-August 1992): 232–233.

63. Root-Bernstein, *Rethinking AIDS*, p. 318.

64. Ibid., p. 317

65. Ibid., p. 320.

66. Ibid., pp. 42–43.

67. Ibid., p. 40.

68. Alan Lifson et al., "HIV Seroconversion in Two Homosexual Men After Receptive Oral Intercourse with Ejaculation: Implications for Counseling Concerning Safe Sexual Practices," *The American Journal of Public Health* 80 (December 1990):1509–1510.

69. "Hunting the Virus Hunter."

70. Gus G. Sermos, *Doctors of Deceit* (McComb, MS: GGS Publishing, 1988), p. 37.

71. "Kimberly's Story," *60 Minutes*, CBS-TV, June 19, 1994.

72. Coleman Jones, "AIDS: Words from the Front," *SPIN*, August 1994, p. 79.

73. Barbara Kantrowitz et al., "Teenagers and AIDS," *Newsweek*, August 3, 1992, p. 45.

74. Frank Richter, "AIDS Threat to Blacks Overblown," *The Detroit News*, January 13, 1989, p. A13.

75. Centers for Disease Control and Prevention, "AIDS Weekly Surveillance Report," November 14, 1988, p. 1.

76. "More Minorities Hit by AIDS," *The Washington Times*, September 9, 1994, p. A3.

77. Michael Fumento, "This Magic Moment," *The American Spectator* (February 1992):18.

78. Ibid., p. 16.

79. J. Philippe Ruston, *Race, Evolution and Behavior* (New Brunswick, NJ: Transaction Publishers, 1995), p. 22.

80. "Hunting the Virus Hunter."

81. Fumento, "Magic Moment," p. 18

82. Ibid.

83. "More Minorities Hit," p. A3.

84. "Is There a Heterosexual AIDS Epidemic?" *Tony Brown's Journal* (1101), January 17, 1988.

85. Michael Fumento, *The Myth of Heterosexual AIDS: How a Tragedy Has Been Distorted by the Media and*

Partisan Politics (Washington, D.C.: Regnery Gateway, 1990), p. 142.

86. "AZT Pregnancy Study. More Questions Than Answers," *New Jersey Women and AIDS, Network News,* Spring 1994, p. 4.
87. "Hunting the Virus Hunter."
88. "Drug May Save AIDS Moms' Babies," *New York Post,* February 21, 1994, p. 9.
89. Neenyah Ostrom, "Duesberg Update," *New York Native,* August 15, 1994, p. 18.
90. John Schwartz, "AIDS Virus Test Urged for All Pregnancies," *The Washington Post,* July 7, 1995, p. A1.
91. Lawrence K. Altman, "Children's AIDS Study Finds AZT Ineffective," *The New York Times,* Feb. 14, 1995, p. C13.
92. Ibid.
93. Ibid.

Chapter 8: The Gang of Frankenstein

1. Deborah Blum, *The Monkey Wars* (New York: Oxford University Press, 1994), p. 229.
2. Ibid., p. 228.
3. "Apocalypse Bug," CNN, May 14, 1994.
4. Centers for Disease Control and Prevention, "Update—Zaire Ebola Outbreak," May 25, 1995, p. 1.
5. Richard Horton, "Infection: The Global Threat," *The New York Review of Books,* April 6, 1995, p. 24.
6. Eva Lee Snead, *Some Call It AIDS, I Call It Murder,* vol. 1 (San Antonio, TX: AUM Publications, 1992), p. 395.

7. Tom Curtis, "The Origin of AIDS," *Rolling Stone*, March 19, 1992, p. 56.

8. Laurie Garrett, *The Coming Plague* (New York: Farrar, Straus & Giroux, 1994), p. 371.

9. Curtis, "Origin of AIDS," p. 57.

10. Ibid.

11. Gina Kolata, "Theory Tying AIDS to Polio Vaccine is Discounted," *The New York Times*, October 23, 1992, p. A16.

12. Ibid.

13. John Seale, "AIDS Virus Infection: Prognosis and Transmission," *Journal of the Royal Society of Medicine* (1985):613–615.

14. "1959 AIDS Case in Doubt," *Poz*, June-July 1995, p. 24.

15. Garrett, *Coming Plague*, p. 380.

16. Snead, *Some Call It AIDS*.

17. Robert Willner, *Deadly Deception* (n.p., Peltec Publishing, 1994), p. 86.

18. Jan Creamer, ed., *Biohazard: The Silent Threat from Biomedical Research and the Creation of AIDS* (London: National Antivivisection Society, 1987), p. 2.

19. Laurie Garrett, "Are We Ready for the Next Plague?" *USA Today*, March 9, 1995, p. A11.

20. Creamer, *Biohazard*, p. 17.

21. Ibid., p. 56.

22. Richard Harris and Jeremy Paxman, *Higher Form of Killing: The Secret of Chemical and Biological Warfare* (New York: Hill and Wang, 1982), p. 240.

23. "Soviets Accuse U.S. of Ethnic Weapons in War of Words," *Chicago Tribune*, June 6, 1987, p. A20.

24. Chet Wade, "Warnings of a Weapons Epidemic," *Pittsburgh Post-Gazette*, January 1, 1989, p. 1.

25. Charles Piller and Keith Yamamotto, *Gene Wars: Military Control over the New Genetic Technologies* (New York: William Morrow, 1988), p. 16.

26. Russell Watson et al., "The 'Winds of Death,' " *Newsweek*, January 16, 1989, p. 24.

27. Ibid.

28. Gerald Horne, "Race Backwards: Genes, Violence, Race, and Genocide," *Covert Action Quarterly*, no. 43 (Winter 1992–1993):30.

29. Lowell Pointe, "Germ Genocide: Shocking New Science Weapons Kill by Race," *Gallery*, October 19, 1978, p. 44.

30. R. Jeffrey Smith, "Iraq Had Program for Germ Warfare," *The Washington Post*, July 6, 1995, p. A1.

31. Horne, "Race Backwards," p. 30.

32. Robert Lederer, "Racially Targeted Biowarfare," *Covert Action Quarterly*, no. 43 (Winter 1992/1993):30.

33. Carl A. Larson, "Ethnic Weapons," *Military Review* 50, no. 11 (November 1970):3–11.

34. William C. Boyd, "Rh and the Races of Man," *Scientific American*, August 1951, p. 22.

35. George Judson, "Scientist Studying Diseases Is Infected by Deadly Virus," *The New York Times*, August 22, 1994, p. B5.

36. L. Fletcher Prouty, *The Secret Team: The CIA and Its Allies in Control of the United States and the World*, 2nd ed. (Costa Mesa, CA: Institute for Historical Review, 1973), p. viii.

37. John Ritter, "Report: Military Knew Risks of Gulf War Drugs," *USA Today*, December 8, 1994, p. 3A.

38. H. Josef Hebert, "Agency Sought Cadavers for Its Radiation Studies," *The Washington Post*, June 22, 1995, p. A 3.

Chapter 9: Team America: The End of Racism and Sexism

1. "Professors Not Impressed with U.S. Students," *The Washington Times*, June 20, 1994, p. A3.
2. William B. Johnston and Arnold H. Packer, *Workforce 2000: Work and Workers for the 21st Century* (Indianapolis: Hudson Institute, Inc., 1987), pp. 85, 89, 95.
3. Henry A. William III, "Beyond the Melting Pot," *Time*, April 9, 1990, p. 28.
4. Ibid.
5. "One on One with Lester Thurow, Part I," *Tony Brown's Journal* (1514), PBS, May 1, 1992, and "One on One with Lester Thurow, Part II," *Tony Brown's Journal* (1515), PBS, May 8, 1992.
6. William Poundstone, *Prisoner's Dilemma* (New York: Doubleday, 1992), p. 216.
7. "Lester Thurow."
8. Peter Annin, "The Corporation: Allstate Saw the Light When It Started Following the Money," *Newsweek*, April 3, 1993, p. 32.
9. Ibid.
10. Heather MacDonald, "The Diversity Industry," *The New Republic*, July 5, 1993, p. 22.
11. Dirk Johnson, "Business Shuns Areas That Look Too White," *The New York Times*, April 18, 1994, p. 8.
12. Ibid.
13. Gary Becker, *Human Capital* (Chicago: University of Chicago Press, 1975), p. 191.
14. "Is the Black Middle Class Angry?" *Tony Brown's Journal* (1705) PBS, February 18, 1994.
15. "One on One with Andrew Hacker," *Tony Brown's Journal* (1518), PBS, May 29, 1992.

16. Norman Mailer, *Advertisements for Myself* (New York: G. P. Putnam's Sons, 1959), p. 337.
17. Ibid., pp. 332–333.

Chapter 10: Political Dynamite

1. David Bernstein, "From the Editor," *Diversity & Division: A Critical Journal of Race and Culture* (Fall 1993):4.
2. DeWayne Wickham, "Black Spokesman Wrong to Join GOP," *Argus Leader*, August 11, 1991.
3. James Harney, "Noted Commentator Joins Republican Ranks," *USA Today*, July 23, 1991, p. A2.
4. GOPAC meeting, audiotape, November 1991.
5. Tony Brown, "Become a Republican," *The Wall Street Journal*, August 5, 1991, p. A14.
6. Charles V. Hamilton, "Promoting Priorities: African-American Political Influence in the 1990s," *The State of Black America* (New York: The National Urban League, 1993), p. 60.
7. "The Devil Made Him Do It," *Rising Tide*, January/February 1995, p. 6.
8. Michael C. Dawson, "Black Discontent: The Preliminary Report on the 1993–1994 National Black Politics Study," NBPS report no. 1, University of Chicago, April 1994, p. 7.
9. Ossie Davis speech, Congressional Black Caucus Dinner, Washington, D.C., June 18, 1971.
10. Ibid.
11. "The Completely Unofficial House Ideological Spectrum," *Roll Call*, sect. B, "Welcome Back Congress," January 24, 1994, pp. 24–26.
12. Dawson, "Black Discontent," p. 1., table 1, B11.

13. David Brock, "Living with the Clintons," *The American Spectator* (January 1994):23.
14. Matthew Rees, *From the Deck to the Sea: Blacks and the Republican Party* (Wakefield, NH: Longwood Academic, 1991), p. x.
15. David A. Bositis, "Meyerson Confused" (letter to the editor), *Policy Review*, no. 69 (Summer 1994):88–89, and HBO/Joint Center Survey, June 1992.
16. David J. Garrow, "On Race, It's Thomas v. an Old Ideal," *The New York Times*, July 2, 1995, p. E5.
17. Ibid.

Chapter 11: A Plan to Make Black America Work

1. "Around the Nation," *Destiny Magazine*, April 1995, p. 13.
2. "Have Capital, Will Flourish: Black Entrepreneurs," *The Economist*, February 27, 1993, p. A33.
3. Harold Cruse, *Plural but Equal: A Critical Study of Blacks and Minorities in America's Plural Society* (New York: William Morrow/Quill, 1987), pp. 79–80.
4. Joel Kotkin, "Why Blacks Are Out of Business," *The Washington Post*, Outlook section, September 7, 1986.
5. "Have Capital," p. A33.
6. Ivan Light, "Ethnic Enterprise in America: Japanese, Chinese, and Blacks," in *From Different Shores*, ed. Ronald Takaki (New York: Oxford University Press, 1987), p. 83.
7. Ibid., p. 89.
8. James M. Washington, ed., *A Testament of Hope: The Essential Writings and Speeches of Martin Luther King, Jr.* (San Francisco: Harper & Row, 1986), p. 648.

9. "The Changing Face of America," *Time*, July 8, 1985, p. 1.
10. Theodore Cross, "Black Africans Now the Most Highly Educated Group in British Society," *The Journal of Blacks in Higher Education* (Spring 1994):92.
11. Ibid.
12. Ibid., p. 93.
13. Ibid.
14. Interview with Solomon J. Herbert, Publisher/Editor-in-Chief, *Black Meetings & Tourism*, June 8, 1995.
15. Barbara Reynolds, "Blacks Should Use Their Billions to Call Their Own Shots," *USA Today*, July 31, 1992, A9.
16. Interview with Girl Scouts of America National Office, Washington, D.C., June 1, 1995.
17. Lena Williams, "Chinese Say Study Hurts Restaurants," *The New York Times*, September 29, 1993, p. C1.
18. Lloyd Gite, "The New Agenda of the Black Church: Economic Development for Black America," *Black Enterprise*, December 1993, p. 56.
19. "Labor Letter: A Special News Report on People and Their Jobs in Offices, Fields, and Factories," *The Wall Street Journal*, November 9, 1993, p. 1.
20. Ibid.
21. Janet Kidd Stewart, "Black Workers Suffer Most as Jobs Move," Chicago *Sun-Times*, Sunday news section, October 31, 1993, p. 15.
22. Joshua Quittner, "How to Send Real Money over the Internet," *Time*, June 12, 1995, p. 64.
23. Lena Williams, "Computer Gap Worries Blacks," *The New York Times*, May 25, 1995, p. C1.
24. Ibid.

Chapter 12: If I Were President, How I Would Make America Work for U.S.

1. "America's New Crusade," NBC/*Wall Street Journal* Poll, *U.S. News & World Report*, August 1, 1994, p. 27.
2. Alfred L. Malabre, Jr., *Beyond Our Means: How Reckless Borrowing Now Threatens to Overwhelm Us* (New York: Vintage Books, 1987), p. xi.
3. Ibid., p. 160.
4. Florence Scovel Schinn, *The Game of Life and How to Play It* (Marina del Rey, CA: DeVorss & Co., 1925), p. 7.
5. Matthew 25:30.
6. Scott Alan Hodge, "Davis-Bacon: Racist Then, Racist Now," *The Wall Street Journal*, June 25, 1990, p. A14.
7. Ibid.
8. Tony Snow, "Enfeebling Davis-Bacon Workplace Fungus," *The Washington Times*, December 13, 1993, p. A19.
9. Ibid.
10. "PACs and the Black Caucus," *The Washington Post*, July 14, 1994, p. A22.
11. Frank McCoy, "Can the Black Caucus Be Bipartisan?" *Black Enterprise*, January 1994, p. 22.
12. Robert Pear, "House Takes Up Legislation to Dismantle Social Programs," *The New York Times*, March 22, 1995, p. A10.
13. Louis Uchitelle, "Union Goal of Equality Fails the Test of Time," *The New York Times*, July 9, 1995, p. A18.
14. David Sands, "Parrel Votes to Repeal Davis-Bacon," *The Washington Times*, March 30, 1995, p. A5.
15. George Reisman, *Capitalism: The Cure for Racism* (La-

guna Hills, CA: The Jefferson School of Philosophy, Economics, and Psychology, 1992), p. 5.

16. Ibid.

17. Ibid, pp. 12–13.

18. Ludwig von Mises, *Socialism* (London: Jonathan Cape, 1936), p. 505.

19. Bettina Bien Graves, *Ludwig von Mises (1991–1973): A Prophet Without Honor in His Own Land* (New York: Freeman, 1995), p. 6.

20. "One on One with Lester Thurow, Part I," *Tony Brown's Journal* (1514), PBS, May 1, 1992, and "One on One with Lester Thurow, Part II," *Tony Brown's Journal* (1515), PBS, May 8, 1992.

21. Eric Pianin, "Panel on Entitlements Starts Its Task Averse to New Taxes," *The Washington Post*, June 14, 1994, p. A4.

22. "Lester Thurow."

23. Stanley H. Brown, "How Can We Tax the Underground Economy?" *New York Newsday*, August 7, 1994, p. A37.

24. Dan Schaefer and Billy Tauzin, "Sales Tax Alternative to Income Tax," *The Washington Times*, February 27, 1995, p. A17.

24. "Eyes on a New Prize," *World*, June 4, 1994.

25. Angela Hulsey, ed., *School Choice Programs: What's Happening in the States* (Washington, D.C.: The Heritage Foundation, 1993), p. 2.

26. Charles Jacob and Mohamed Athie, "Bought and Sold," *The New York Times*, July 13, 1994.

27. David Brinkley, *This Week with David Brinkley*, ABC-TV, February 4, 1995.

INDEX

─── F R E E ! ───

One-month subscription for up to 10 hours
of *Tony Brown Online* core services with Internet access
<u>a $9.95 Value</u>
All you need is a computer, a modem, and our FREE software.
Simply clip and mail this original form.

THIS FREE OFFER INCLUDES:*

1. **FREE software** for *Tony Brown Online*, a membership-based computer network service to be launched nationwide Thanksgiving weekend 1995
2. **One FREE month** of *Tony Brown Online*'s core services with Internet access, a $9.95 value
3. **Up to 10 FREE hours** to explore the Internet**

*A $10 one-time set-up fee and a credit card, checking or debit account, or some other autho-rized form of billing to register for the free-month demonstration are required. You will not be billed unless you sign up for a permanent account. The number of hours vary, depending on long-distance charges. Limited to one free account per household. New members only, please.
**Up to 5 hours outside the New York City area.*

THE EMPOWERMENT NETWORK

In addition to linking subscribers to the estimated 30 to 50 million people world-wide using the Internet and providing a variety of online entertainment, the new membership-based computer network, *Tony Brown Online*, is designed to meet today's challenges of technological displacement of workers, economic uncertainty, single-parent homes, and computer-deprived schools. "We cannot take every young-ster from poverty to the middle class through Harvard. But we could take every youngster from poverty to the middle class with a computer" is the way Tony Brown describes his human capital approach to self-sufficiency.

─── *ACT NOW, LIMITED OFFER* ───

Clip and mail this original form to:
TONY BROWN ONLINE
1501 BROADWAY–SUITE 412
NEW YORK, NY 10036

(please print)

NAME: _____

BUSINESS: _____

MAILING ADDRESS: _____

CITY: _____ STATE: _____ ZIP: _____

PHONE: ( ) _____ FAX: ( ) _____

Information:
VOICE: (212) 575-0876   FAX: (212) 391-4607
EMAIL: tonybrown.com   WORLD WIDE WEB: http://www.tonybrown.com

This offer is made by Tony Brown Productions, Inc.
William Morrow assumes no liability or obligation whatsoever with respect thereto.